THIS VITALLY IMPORTANT DOCUMENT IS
COMPILED FROM EXPERT TESTIMONY,
SCIENTIFIC STUDIES, GOVERNMENT
INQUIRY AND THE GROWING BODY OF
DATA IN THE FIELD. ITS PURPOSE IS TO
INFORM THE PUBLIC OF THE TRUE FACTS
ABOUT A TOPIC OFTEN CLOUDED BY
FICTION, SUPERSTITION, AND ALARMIST
MISREPRESENTATION.

THE WEATHER CONSPIRACY
The Coming of the New Ice Age
is A SPECIAL IMPACT TEAM REPORT

Investigative reporters: Pete Kilroy, Becky Lazear,
Sandy Porter
Writers: Alastair Clark, Tessa Clark, Penny Naylor,
David Runnalls, Peter Verstappen
Background Specialists: Drayton Bird, Jay Callen,
Jack Porter, Mo Ramm
Back-up: Jean Sweeney Campbell, B.E.A. Draper,
Mary Hedges, Margaret Ninker, Karen Sturges,
Sue Symes

THE
WEATHER
CONSPIRACY

The Coming of the
New Ice Age

a report by
THE IMPACT TEAM

Heron House
Publishing International Ltd.

BALLANTINE BOOKS • NEW YORK

Library of Congress Catalog Card Number: 77-5928

ISBN 0-345-27209-9

Manufactured in the United States of America

First Edition: April 1977

Contents

Introduction

The function of research within the Agency has been
directed at defining the relationship of climatology to
the intelligence problem. It is increasingly evident that
the Intelligence Community must understand the magnitude of international threats which occur as a function
of climatic change. These methodologies are necessary
to forewarn us of the economic and political collapse
of nations caused by a world-wide failure in food production. In addition, methodologies are also necessary
to project and assess a nation's propensity to initiate
militarily large-scale migrations of their people as has
been the case for the last 4,000 years.

—CIA report

The weather is always with us. We take pleasure
in good days and make the best of bad ones. And there
is always "the weather" to fill awkward gaps in con-
versation. But most of all we take it for granted. We
know that for much of the Northern Hemisphere the
year is divided into four distinct seasons: winter, with
its temporary snow; spring, which brings lilacs and
daffodils; the halcyon days of summer; and autumn,
with its crisp, clear days.

In part we take the weather for granted because it is
seldom really noticeable. For the past fifty years we
have enjoyed some of the best weather mankind has

ever known. But this exceptional clement period has lulled man into a false sense of security. Now, according to many climatologists and some recently declassified Central Intelligence Agency reports, our climate and the consistency of our seasons can no longer be taken for granted. Instead, we are in for major climatic change.

In recent years, the CIA has come under public scrutiny. Near-daily headlines have revealed scandals as various as the overthrow of Allende in Chile and conspiracies to spy on Americans abroad. There have been abuses of power. But, incidental infamy aside, the CIA is the cornerstone of the U.S. intelligence community, and as such has the responsibility of assessing the impact of changing world conditions on the United States, its foreign policy, and its defense. Its analysts are a group of experts who take a cold, dispassionate view of world events today and their ramifications for America tomorrow.

In the early 1970s top CIA thinkers concluded that changing weather was "perhaps the greatest single challenge that America will face in coming years." As a result, they ordered several studies of the world's climate, the likely changes to come, and their probable effect on America and the rest of the world. The studies conclude that the world is entering a difficult period during which major climatic change is apt to occur.

The ramifications of such a change would be many, affecting the foreign policy and your fuel bill, what food is grown and what you eat. Taken at its bleakest, the coming weather may signify massive migration and equally massive starvation. Taken at its most optimistic, it will cause all of us to change parts of our lifestyle.

We on the Impact Team have attempted to deal with this complex, contradictory, and confusing topic in such a way as to present a realistic insight into what has occurred today, and an intelligent appraisal of what the future may hold. The facts presented and the conclusions drawn speak for themselves. They show quite clearly that while different climatologists forecast different futures, there is general agreement that weather will now "conspire" against humanity and the world we've made. If changing weather patterns are not to result in unimaginable famine and plague, how we think, act, and eat must change—starting right now.

1

Normal Weather Is Back—The Coming of the New Ice Age

The weather we call normal is in fact highly abnormal . . . There is growing consensus among leading climatologists that the world is undergoing a cooling trend . . . excellent historical evidence exists from areas on the European plains . . . [researchers hypothesize that] the change from an interglacial to glacial time period could take place in less than 200 years.

—CIA report

"Every year we've prayed for snow, but this is ridiculous!"
—Upper New York State ski-operator, January, 1977

The winter of 1977 was something else again. Unprecedented in ferocity, it seemed to go on forever. The average American in the upper Midwest and the Northeast followed the wind-chill factor more avidly than the sports statistics. After all, it was right outside the door. While that part of our country was snowed under, Aspen's skiers were looking at bare rocks, Floridians were watching their crops freeze, Californians went surfing (possibly in an effort to forget their drought), and Alaska's kodiak bears were wishing someone would turn off the sunlamp so they could go to bed—the thermometer stayed in the forties throughout most of January.

For two thirds of the United States it was the coldest winter on record. Upper New York State was buried under twenty-six feet of snow. Buffalo lay inert as a fifteen-foot fall brought life to a complete standstill for its 1.3 million citizens. A freak frost in Florida reduced the citrus crop by 14 percent from the previous year. It didn't stay long—Tallahassee recorded 17°F for just a few hours—but its icy fingers caused losses estimated at $125 million. Snow fell briefly in Miami, and caused the Miami *Herald* to headline: OLD MAN WINTER LOST ALL HIS ROAD MAPS.

Chicago, Pittsburgh, Jacksonville, and Nashville had many days of the lowest temperatures on record. Cincinnati averaged 20° below normal; Indianapolis shivered for weeks and recorded an all-time low of 18° below. The Ohio River was blocked by ice, while cargo ships in and out of Baltimore and Delaware moved in cautious convoys led by icebreaking tugs.

The Chesapeake was choked with up to five feet of ice, and its fabled crab fishermen spent their winter days crabbing in unemployment lines. Lake Erie assumed the eerie fastness of the world's largest skating pond.

At least seventy-five deaths were attributed to snow, and to the deadly wind-chill factor, which can freeze a person in his tracks—forever.

Abnormal weather, people said. And their belief that the winter was altogether too bizarre to happen again in their lifetime was reinforced by news from the far West. Folsom Lake reservoir, near Sacramento, California, was down to forty feet below normal water level as parts of the state sweltered under a severe drought. Many towns in the north resorted to water rationing. Marin County Water Authority spokesman J. Dietrich Stroeh called the situation "frightening."

"The problem is worse than anyone realizes," said Daryl Arnold of the Western Growers Association, as reports came in that many square miles of California farm land were already too dry to grow anything. The minimum estimate of crop damage in California (where 40 percent of the nation's fruit and vegetables are grown) is $2 billion. When the ripple effect is taken into account (unemployed agricultural workers don't eat out, buy new cars, etc.), a conservative estimate of the loss to the state's economy is $5 billion.

Thirteen other states, from Washington and Oregon across to Kansas and Nebraska, were also hit. Wheat and oat stocks are depleted. In parts of Minnesota, the Dakotas, Wisconsin, and Kansas the earth froze to a depth of four feet, because not enough snow fell to give the usual insulation to the ground. Farmers in all these states are worried about this year's beef and wheat yields.

Twenty-four skiers in the state of Washington emphasized the winter's topsy-turvy weather when they donned tuxedos and dined alfresco five thousand feet up Mount Washington—on dry ground. Normally their dining table would have rested on eighty inches of snow, and they would have worn mountaineering gear. But then, parts of Alaska were, however briefly, warmer than Florida. However, America's winter of discontent is not the only recent sign of a weather system running amok.

Old Man Winter was perplexing other countries as well. In the U.S.S.R., snow fell farther south than in any winter on record. Neighboring Hungary and Czechoslovakia were flooded out with a freak warm spell. Britain had severe blizzards in random parts of the country. Several parts of southern Australia were declared disaster areas as the worst bush fires ever torched timber and crops. In Brazil, the São Francisco River, an important commercial artery, became too shallow along most of its eighteen-hundred miles for cargo boats to pass through.

And, in the not-so-distant past, there have been other startling weather-related events. In 1972 Russia's total wheat crop failed. Since 1970, 3 million have died as a result of drought in the six countries of the African Sahel. The once-teeming schools of anchovy off the coast of Peru have mysteriously disappeared. In India 150 million people would have starved had help not come in time when 1974's monsoon failed to show.

If you spent the summer of 1976 in damp olde England, you experienced the hottest and driest weather ever recorded there; if you drove the California coast, you passed a desert in the making. And if you happened to hear from friends in Melbourne, Australia, in early February of 1977, the news was hot—it hit 120°.

No doubt, there was something else in your mailbox that made you realize the weather was suddenly creating a major impact on your life: a skyrocketing utility bill (or none at all if you'd been turned off).

And when weather acts up, other acts follow. You don't have to be the world's best economist to know that scarcity creates demand and demand causes price rises. U.S. fruits and vegetables will cost more throughout 1977. If a city has to spend unplanned millions on snow removal, it's got to get the money from somewhere. That's called your taxes, and, given the rise in unemployment, they will be drawn from a smaller tax base.

To no one's surprise, then, the subject of weather in 1976–77 moved from a small box at the back of the newspaper to the front page, from the tail end of the news on TV to the lead item on *CBS News*, with Walter Cronkite talking about fifteen feet of snow in Buffalo and more to come. And more will come. You don't need a crystal ball to predict that when five times the normal amount of snow melts, spring flood waters will rise faster than winter fuel bills.

At the same time, the Midwest, America's breadbasket, has too little moisture, which could turn it into a totally unproductive dust bowl. Thirteen western states, with only one fifth their normal mountain snow cover, can look forward to a bleak period of water rationing and power blackouts if summer, their dry season, follows its usual course.

The CIA's Judgment of the Weather Situation

The newspaper headlines read, FREAK STORMS . . . WORST WEATHER EVER . . . ALL-TIME LOW. Stories often went on to say it couldn't happen again. But it very definitely can. That is the consensus of recently released Central Intelligence Agency reports, which highlight the fact that we are overdue for a new ice age.

What's going on?

The answer is simple. America and the rest of the Northern Hemisphere are returning to normal weather. After fifty years of unprecedentedly temperate climate, the world is returning to the more radical shifts and the cooler climate that characterized the previous eight hundred years.

Many climatologists believe that since the sixties the world has been slipping toward a new ice age. The only questions in their minds are: What kind of ice age will it be, little or great? How soon will it happen?

As intelligence analyses, the CIA studies of the effects of the weather to come on the world's geopolitical balance are not intended to plead a case or take a biased point of view. Instead, they form a coldly rational judgment of our current situation. The studies have profound and disturbing implications: "The politics of food will become the central issue of every government in the near future." This estimate is based on overwhelming evidence gained in the past two decades that the world is heading into an unstable weather future—what climatologists call the "cooling trend." Using satellites, computers, ice cores, carbon-dating, and other advanced techniques, climatologists today have a far clearer idea of climate patterns than ever before.

Reid A. Bryson, of the University of Wisconsin, is one of the world's most renowned and respected climatologists. Professor Bryson makes the point that for fifty of the past sixty years, the world's climate has been *abnormal*, not normal. He convincingly demonstrates that the world has enjoyed the best agricultural climate since the eleventh century.

During this fifty-year period, and notably during the 1960s, a worldwide "Green Revolution" occurred. Through the use of highly selective, specially created strains of rice, wheat, and other grain crops, famine in underdeveloped countries was pretty well staved off. At the same time, the U.S. produced bumper crops. Monsoons arrived on schedule in India, West Africa, and China. By 1955, alpine glaciers had retreated by a half mile when compared with their 1865 limits. The globe was warm and its people were fairly well fed. Other experts' analyses agree with Bryson's.

Hubert H. Lamb, director of the Climatic Research Unit at the University of East Anglia, England, is one such expert, who analyzes the weather's past to predict its future. Professor Lamb bases his statistics on ships' logs, manuscripts, legends, and the accurate and comprehensive records kept in England since the late seventeenth century. Cesare Emiliani, of Miami University, reaches similar conclusions. He calculates that over the past 7 million years the global mean temperature has been as high as it is in this century only 5 percent of the time.

Experts disagree as to when this exceptionally favorable period of abnormally good weather began to end. Some say it was as early as the late forties; others say it was as late as the mid-sixties. But they are in general agreement that our weather, the single most important factor in civilization's survival, is now changing—and for the worse.

Since the sixties, changes in the weather pattern have been examined with special care. The results of this examination have been disturbing.

The ice cover in the Northern Hemisphere increased by 12 percent in 1971—an increase equal to the combined area of England, Italy, and France. This added ice has remained.

The great ice mass of Antarctica grew by 10 percent in one year, 1966–67. And the average ice and snow cover is high—and expanding.

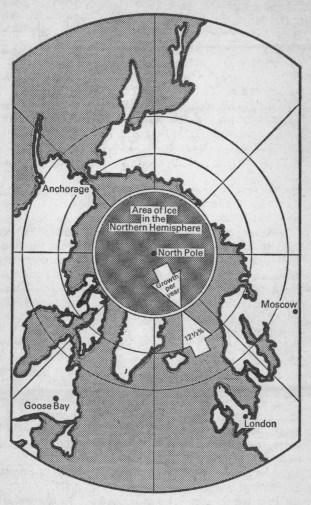

Annual expansion of sea ice

Satellite photography has captured another, equally disturbing picture: From 1967 through 1973, winters in the Northern Hemisphere have grown longer by almost one month, averaging 84 days in 1967 and 104 days in 1973. The inhabitants of Greenland would confirm this. Greenland suffered below-normal temperatures for nineteen solid months in the early 1970s, the severest cold in over a hundred years.

And as the evidence accumulated, the world's climatologists turned their thoughts to the unthinkable: the real possibility that the world is about to undergo a major change in weather—and face a new ice age.

What Kind of Ice Age?

Ice ages are cyclic. In the past seven hundred thousand years the world has suffered eight great ice ages. We know their general pattern: Each lasts about ninety thousand years and is followed by a warm period of about ten thousand to twelve thousand years. These warm years are known as the interglacial period; the earth appears to be emerging now from the most recent of these.

Nature provides man with early-warning signals of the coming cold. Scientists have long been interested in the migratory patterns of birds and animals; they have proved that variations from normal patterns can have significant implications for man. At present, it is the armadillo that is sounding the alarm. Between 1880 and 1940, the warmest period in the last eight hundred years, the tropical armadillo invaded the U.S. and migrated as far north as Nebraska. Since 1940, armadillos have steadily retreated south. Similarly, in Europe a species of warmth-loving snails became extinct during the 1960s. Such seemingly insignificant natural signs are harbingers of major change.

What kind of change will it be?

LITTLE ICE AGE

Professor Bryson feels that the evidence suggests the change will be a return to the climate that was dominant from the seventeenth century to about 1850. During most of that time northern Europe lived in the twilight of permanent winter. Malnutrition caused great plagues in Europe, Russia, India, and Africa, while the Iberian peninsula, Italy, and Greece thrived.

In northern Europe, even the summers were cool and dry, and harvests were poor. The winters were snowy and bitter cold. Evidence of just how cold it was can be seen in the fact that London's River Thames froze during severe winters—and so solidly that the city's population held fairs on its ice that featured the roasting of enormous oxen.

The CIA has analyzed Bryson's and others' conclusions and believes that "a global climatic change is taking place" and that "we will not soon return to the climatic patterns of the recent past." One Agency report (see Appendix 1) quotes "The Wisconsin Study" (the University of Wisconsin was the first accredited academic center to forecast a major global change): "The Wisconsin forecast suggests that the world is returning to the climatic regime that existed from the 1600s to the 1850s, normally called the neo-boreal, or 'Little Ice Age,' " and add that the political, historical, and economic consequences of that climatic era have been masked up until now by the historian's preoccupation with technical progress.

If we are about to enter a new Little Ice Age, what will it bring? Professor Bryson has put forward a likely scenario of what could happen, based on events during the nineteenth century, part of the Little Ice Age:

More rain would fall in the northern U.S.

The central Gulf coast, the Southwest, and the Rockies would become drier.

The winter-wheat region of the high Plains would be much wetter, although yields would not be affected.

But his prognosis for the rest of the world is much gloomier:

India would have a major drought every four years, and 30–50 million metric tons would be required from the world's grain reserves to prevent the death of 150 million Indians.

China would have starvation conditions every five years and need 50 million metric tons of grain for its people.

Canada would lose 50 percent of its food production, and reduce exports by 75 percent.

Professor Bryson's reasoning is as follows: Up to about 1940, the world warmed; then, from about the mid-forties on, a reversal took place as polar air expanded south and the Northern Hemisphere cooled. He believes that the three main factors involved in these trends are volcanic dust, man-made dust, and carbon dioxide.

Dust filters the heat and light of the sun. Early in this century, volcanic activity practically stopped; thus, more of the sun's radiation came through. But, says Bryson, "there are potfuls of volcanoes erupting away now."

We can't control volcanos—and man-made dust has increased rapidly in the past few decades. One CIA report calculates that today it makes up about 30 percent of all the dust in the atmosphere. Dust cools the polar regions more than it does the tropics. The final effect is cooler weather in the northern atmosphere, close to the Arctic air masses.

Iceland is used as a barometer of climatic change because of its location, close to the Arctic Circle. In the past thirty years the temperature there has dropped by about 1°F compared with the preceding thirty years. In the same period, the average temperature over the Northern Hemisphere went down by 0.9°F.

The seemingly small difference in temperature is vastly significant. Twenty-three thousand years ago, in the last Great Ice Age, giant icebergs appeared as far south as today's Mexico City—and the Earth's annual mean temperature was only 7.2°F cooler than it is today.

Other climatologists look beyond Bryson's scenario to a giant freeze. Soviet weatherman Mikhail Budyko believes that a 2.8°F drop in the average world temperature would start the glaciers on the march. If the temperature should fall by another 0.7°F, it could usher in a ninety-thousand-year tyranny of ice and snow.

GREAT ICE AGE

What would a Great Ice Age be like? The last Great Ice Age, which ended about ten thousand years ago, engulfed New York and came down as far as Columbus, Ohio. Chicago was buried under ice a mile thick. The one before that came beyond where St. Louis stands today. Ice Age glaciers created Cape Cod and Long Island, and carved out the Great Lakes.

Most of Canada, southern Alaska, and the northern part of the United States were covered by a single ice sheet the size of Antarctica. It merged with the ice on the Rockies and, as a series of glaciers, stretched down through California to Mexico. It was fringed by a narrow strip of tundra about one hundred miles wide that separated the ice from the spruce forest that covered vast areas of America.

In Europe, ice sheets stretched from Scandinavia to Britain, and east far into Russia. Farther south, the Swiss Alps were one gigantic lump of ice. England, Holland, and Germany were icy deserts. The rest was a continent without trees, except in southern France. Where vines grow today, the land was given over to Lapland birches. In an ice age the area around the equator becomes dry, with short, sharp bursts of heavy rainfall.

As water became locked into the ice on the continents, new landmasses rose out of the sea. Thus, the Bering Strait became a bridge between Siberia and

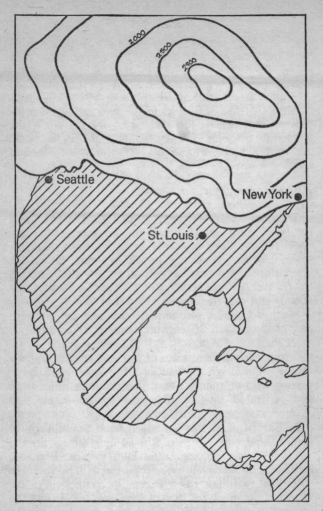

Great Ice Age map of North America. (CLIMAP, International Decade of Ocean Exploration, National Science Foundation. Copyright © 1976 by the American Association for the Advancement of Science.)

Alaska during the last Great Ice Age and cut the Arctic Ocean off from the Pacific.

In the Southern Hemisphere southeast Asia, between China and Java, grew larger. Parts of Australia's now-arid outback were cool and rain-swept. Glaciers reached New Zealand and carved out South America's Andes mountains.

Ocean currents changed direction. The Gulf Stream, for example, headed toward Morocco, and the sea just north of Spain was just 35.6°F—the temperature of the waters around Greenland's sea ice today.

How Soon Would It Happen?

Scientists disagree as to how quickly the change to an ice age—little or great—would occur. One school of thought claims it will be a slow change over many thousands of years, which would give man plenty of time to adapt. A second school believes that the change could occur within ten years.

THE SLOW-FREEZE THEORY

According to this theory, ice packs form from the snow that falls on the tops of mountains. Small icecaps gradually appear, and the cold air that surrounds them encourages more snow to fall. The more the snow builds up, the more it encourages new snowfalls to freeze instead of melt. The sheet grows thicker and moves outward. The ice grows from within.

Another theory holds that the ice packs come down from the Arctic north, slowly moving into warmer regions to the south. This creates a zone at the pack's edges which is hospitable to fresh snowfalls. These, in turn, slow the rate of the pack's southward movement.

Both processes would take thousands of years.

THE FAST-FREEZE THEORY

This climatic theory goes under a frightening name: snowblitz. The snowblitz is a blitzkrieg of snow that in

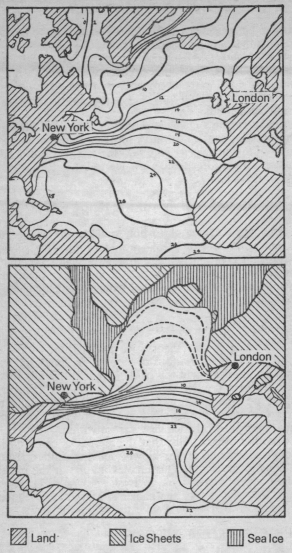

Gulf Stream and sea temperatures today (top) and 18,000 years ago. (B. Bolin.)

| | Land | | Ice Sheets | | Sea Ice |

a few brief years could turn east Canada, northern Europe, the Alps, and other northern regions into ice sheets and set off the inexorable march southward of the world's many glaciers. Like paratroopers, these snowflakes fall from the gray sky and seize the land. But, unlike paratroopers, they cannot be dislodged. Snowblitz flakes do not entirely melt.

Instead, they lie throughout the summers, reflecting the sun's heat with their blank whiteness. Fresh snow bounces up to 85 percent of sunlight back into space. The air stays cold, which means that if succeeding winters bring more snow, the flakes will lie. This process goes on until the land is shrouded in ice all year round. And this ice sheet only has to reach a thickness of twelve inches to make a new major ice age irreversible. It is believed that the process could take only seven or ten years.

Last winter produced very heavy snowfalls in the Canadian Rockies and the Alps, and also in east Canada, site of a great ice sheet in the last ice age.

There is some evidence to support a sudden-change theory. In the U.S. and the U.S.S.R., mastodons have been dug up, perfectly preserved in their coffins of ice. They were standing upright, with partially digested grass in their stomachs. Some people believe that only a snowstorm with the ferocity of a snowblitz forerunner could have frozen them so swiftly in their tracks.

"A unique aspect of the Wisconsin analysis," says one CIA report (see Appendix 1), "was their estimate of the duration of this climatic change. An analysis by J. E. Kutzbach (Wisconsin) on the rate of climatic changes during the preceding 1600 years indicates an ominous consistency in the rate at which the change takes place. The maximum temperature drop normally occurred within 40 years of inception. The earliest return occurred within 70 years. The longest period noted was 180 years."

In Greenland, Willi Dansgaard and his team from Copenhagen University drilled deep into the ice pack for samples of ancient ice. Abundant oxygen 18 in a sample indicates warmth; a sparse quantity means

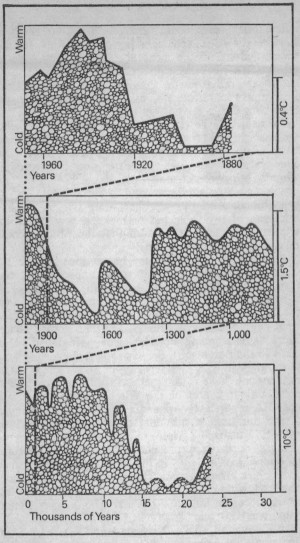

Temperature changes over 80, 900, and 30,000 years.
(B. Bolin.)

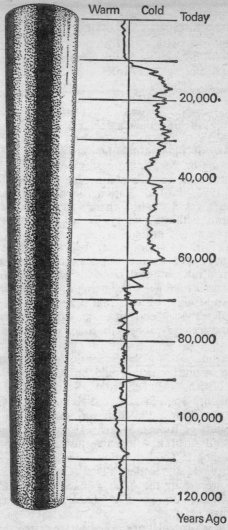

Warm　Cold　Today

20,000.

40,000

60,000

80,000

100,000

120,000

Years Ago

Ice-core findings. (redrawn from a *National Geographic* illustration.)

cold. Professor Dansgaard found that about nine hundred thousand years ago Greenland was basking in a spell of warm weather. With dramatic swiftness—in less than a century—the world apparently plunged into bitterly cold weather. And, as the oxygen 18 level showed, it took a thousand years for the earth to return to warm weather. Dansgaard has also shown that the oxygen 18 in Greenland's ice has been diminishing since 1930.

U.S. fossil hunters looking at samples from the ocean floor off Mexico found evidence of an abrupt change to cold weather about ninety thousand years ago. In 1975, the U.S. National Academy of Sciences said there was a "finite" chance that a new ice age could occur within a century. British scientists estimate the chance as one in ten.

British writer Nigel Calder believes that the new snow age is upon us, that our ten-thousand-year warm period has run out, that the odds are only twenty to one *against* an ice age's beginning in the next hundred years. And Calder believes that if we assume a sudden thousand-year-long cooling as an extra risk over and above the ice age proper, the odds shorten to ten to one.

According to his global scenario for the next Great Ice Age, countries that would be covered by ice include the Irish Republic, Britain, Scandinavia, Finland, Switzerland, Nepal, Sikkim, and New Zealand. Glaciers would once again grind their way through the United States, parts of Russia, Mexico, Holland, Afghanistan, China, and Australia—to name just a few regions.

According to Calder, following a listing by Rhodes Fairbridge, Mali, Upper Volta, Ghana, Togo, and Nigeria are among the African states that would suffer severe famine. So would Afghanistan, Pakistan, and Indochina.

If Calder's calculations are accurate, more than half the earth's inhabitants could die of hunger, and more than a dozen countries could be wiped off the face of the earth.

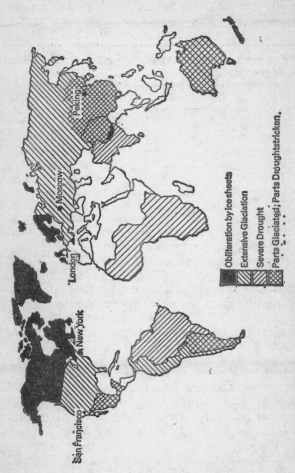

San Francisco

New York

London

Moscow

Peking

Obliteration by ice sheets

Extensive Glaciation

Severe Drought

Parts Glaciated; Parts Droughtstricken.

(Nigel Calder/BBC Publications.)

There is evidence for an advance into cooler weather —a Little or a Great Ice Age. Just how long the process, if uninterrupted, would take, is not known. Scientific advances hold out hope that within a few years future weather will be forecast with a greater degree of certainty. Meanwhile, there is the winter of 1976–77 to remind us that the fickle ferocity of the climate can make a mockery of mankind's efforts and activities— in a matter of days.

2

The Uneasy Marriage—Climate and Civilization

Recently, some archaeologists and historians have been revising old theories about the fall of numerous elaborate and powerful civilizations of the past, such as the Indus, the Hittites, the Mycenaean, and the Mali empire of Africa. There is considerable evidence that these empires may have been undone not by barbarian invaders but by climatic change.

—CIA report

Great civilizations in the past have almost invariably owed their rise to favorable climatic and geographical situations. The warm and fertile river valleys of the Nile, the Indus, and the Euphrates bear witness to this fact. And for centuries it has been the received wisdom that such civilizations, and the mighty empires they produced, died because they were overwhelmed by either internal strife or barbarian invasion—and sometimes a combination of both.

Today, a startling new possibility is being considered —that many of the elaborate and powerful empires of the past were perhaps undone as much by climatic change as by external attack or internal decay.

Man has surely always been, and is, affected by

climatic change; to what degree is an attractive topic for speculation, although not for the purposes of this discussion. But what of his effect *on,* his contribution *to,* that change? It is not necessary to theorize on events in the distant past in order to confirm that there is a mutual relationship between man's activities and environmental shifts. We've only to look back to the recent past.

Dust to Dust

In the 1930s, a decade-long drought devastated the high plains of the southwestern Mississippi Valley. Originally, tall buffalo grass had grown there, supporting countless herds of buffalo.

Dry spells in the mid-nineteenth century partially dried out the area (probably contributing as much to the near-extermination of the buffalo as the trigger-happy hunters riding through on the new railroad). After the buffalo were virtually gone, the pioneers claimed the land. They fenced off their property and plowed up the remaining buffalo grass. In less than a century, the widespread prairies had been transformed into a mosaic of corn and wheat fields.

The drought hit at the start of the Great Depression. The farmers hung on doggedly, sowing and reaping ever-decreasing yields. Nothing in the advanced American farm technology could solve the problem. The soil got drier and drier. The drought intensified. Finally, nothing was left but clouds of dust billowing over the prairies, and the stark silhouettes of leafless trees.

In less than ten years, the Dust Bowl had been created. Thousands of families abandoned their run-down farms and migrated to California. Many were not far from starvation.

As our civilization has grown more and more sophisticated, so it has grown more and more destructive. Today what we do affects our environment more profoundly than ever before. And more swiftly.

Our modern civilization—technologically advanced beyond the wildest dreams of the ancient grain-based civilizations—can inflict climatic damage on a scale and with a speed nothing short of horrifying. The leveling of vast forests, the overgrazing of hillsides, the burning of irreplaceable fossil fuels, the racing of jets through the stratosphere are all affecting our weather and environment in a way that threatens the very basis of our civilization.

The Migration Solution

From Neolithic times until a mere century ago, man always had a solution to climatic change: He migrated.

He depleted game populations, cut forests, stripped fields, dug, sowed, reaped, irrigated, and broke down nature's equilibrium. Then the weather changed. The rivers and wells dried up. The animals died or moved off. The plants no longer grew. And man moved on—to new frontiers, new plains, new valleys, new forests, new rivers. The cycle of waste began all over again.

At the end of the last ice age, Cro-Magnon hunters moved into the valleys of northern Europe. The glacial ice was retreating and fresh plants had sprung up in the newly exposed tundra to nourish the dwindling herds of reindeer and other game.

During the same glacial age, when the ice was thick and the oceans were low, hunters of Mongolian stock migrated from Siberia to Alaska across the land bridge that spanned the Bering Strait.

It was during periods of unfavorable weather that Abraham led the Jews out of Mesopotamia into Egypt; Moses led them back to the Levant when drought blighted once-green areas of North Africa and Arabia. Hittites descended on Babylonia; Medes swept into Assyria. And the glittering Mycenaean civilization fell to Illyrian invaders marching down the Greek peninsula.

Such massive disruptions of population were equally

common in the New World. High on the central Mexican plateau, a chain of broad, shallow lakes made agriculture possible for many of the ancient Mexican civilizations. But when drought hit the valley, the lakes dried out. Whole populations were forced to migrate. The Archaic civilizations collapsed during severe drought in 500 B.C. The Toltecs were driven from Tollan (now Tula) in the twelfth century A.D. And for hundreds of years the Aztecs were homeless nomads, on the move from one drought-seared land to another, until they founded their city of Tenochtitlan in about A.D. 1395.

Two centuries later, Cortés—himself a migrant from arid Spain—ended the Aztec culture. Today, the gleaming pyramids of Tenochtitlan are just a memory in the Aztec codices. The dusty streets of Mexico City, where eight million pairs of feet tread every day, have replaced it. And Lake Texcoco, where Montezuma once boated through fragrant floating gardens, is a forgotten dream.

No longer are there fresh grazing lands hidden behind distant mountains, or fertile valleys tucked away on undiscovered continents, waiting to pour forth their riches, as the people of sub-Saharan Africa recently learned.

The Sahara has sometimes been a green and fertile region. The paths of forgotten rivers coursing south into equatorial Africa are still visible to anyone flying over the desert. Waves of sand roll southward, washing relentlessly against the grasslands at the desert's edge—the Sahel, the "shore" of the desert. Twenty-four million people live there. Even in the best of times they barely eke out a subsistence from the hostile land. But —with luck—they have managed to survive.

In 1968, their luck ran out. The monsoon winds that bring the sparse annual rainfall to West Africa failed to appear. Not just that year, but for six years in succession. It was one of the worst droughts in living memory. Millions of acres were turned into desert.

Thirty to 70 percent of the livestock, on which the

inhabitants depend, perished of thirst or starvation. Crumpled bags of skin that had once been cattle, camels, and goats marred the sunbaked earth, their bones picked clean by well-fed vultures.

Some 10 million starving tribesmen fared almost as badly as their herds. The herdsmen were first forced back to the farm land, crowding into fields that could support no flocks, seeking shade beneath trees long bare of leaves. Then they migrated farther—into towns without water, into refugee centers without adequate food or medicine.

But there was no escape. To the north of the Sahel lay the parched wastelands of the central Sahara. To the south were the teeming nations of West Africa, already suffering from serious overpopulation problems, unable to offer succor to the refugees.

The fate of the Sahel may well serve as a lesson for the future of mankind.

Nature's Answer

Robert Ardrey puts forward the idea that ever since man ceased to be a hunter and began to rely on agriculture, he has been enmeshed in a biological trap. After he learned to grow grain in excess of what he needed, his numbers multiplied to match the available stores of food. More food was then required, so he worked our fields harder, producing ever-more-abundant crops.

Then man was forced to look for new lands to cultivate. And the circle carried on, repeating itself endlessly—until now, when there are no new lands left, and when there are at least two thousand times more people to feed as there were one hundred centuries ago, when the cultivation of grain was a brilliant innovation.

Nature has a way of dealing with the problem of imbalance between food and population, land and weather. She unleashes her most deadly and effective weapons. "In the past year," reported the CIA in 1974, "data from the Sahel, Ethiopia and India indicate that for each death caused by starvation, ten people died of epidemic diseases such as smallpox, and cholera. Bodies weak from hunger are easy prey to normal pathogenic enemies of man."

Famine and epidemic travel side by side. When the starving millions of the Sahel huddled together in ill-equipped, overcrowded refugee camps, infectious diseases and parasites thrived in their undernourished bodies. Infants, young children, and old people were the first to die. Smallpox, typhoid, diphtheria—even measles—took their ghastly toll. People dropped like abandoned livestock on the baked earth, their pathetic corpses often left untouched. No one had the strength to bury them.

Of all the countries of sub-Saharan Africa, perhaps none suffered quite so hideously as Ethiopia, on the easternmost fringe of the monsoon belt. In the northern province of Eritrea, and in the west along the border with the Sudan, as many as four thousand people died every week at the height of the drought in 1973.

Incredibly, the government of Ethiopia tried to conceal the extent of the disaster from its own citizens and the rest of the world. Grain reserves stored in the capital of Addis Ababa were not sent north to feed the starving peasants. It was not until the autumn of 1973, after months of intolerable drought, that the government finally asked the world for help. By then, over one hundred thousand people had died. Although the *immediate* causes of death were cholera, pneumonia, gastroenteritis, and other diseases, the *root* cause was the massive starvation brought on by the failure of the monsoon.

One of the most dramatic consequences was the collapse of the longest reign in modern times. Ethiopia's monarch, Haile Selassie, was unceremoniously dethroned after forty-four years.

It is no accident that the great famines of the past century occurred in countries like China and India, where the ecological balance has been most overstrained. Indeed, there is a strong possibility that the Western world's survival with so few mishaps is entirely due to a happy climatic freak. And it is no mere chance that the great plagues and famines of Europe ended in the 1850s—when the Little Ice Age petered out.

The Little Ice Age

Scholars disagree as to the exact beginning of the Little Ice Age—1250, 1320, 1430, and 1660 are all given as starting dates. But most agree that it ended about 1850. Whatever the dates, the consequences for humanity were devastating. The centuries before the Little Ice Age had been warm. About A.D. 800 the temperature in Europe had taken a turn for the better. Areas that for centuries had barely thawed out from one winter to the next started experiencing balmy summers and longer growing seasons.

Scandinavians, spurred on by this new mildness, and the sudden bountiful food supplies, set off in their sturdy longboats to explore half the globe. They invaded Russia, Britain, and France. They ventured southeast to the Black and Caspian Seas. They passed through the Bosphorus and reached Constantinople. They rowed down the west coast of Africa at least as far as the Canary Islands. And they sailed west along the southern coast of Greenland, through seas that had been packed with ice just a few centuries before.

In A.D. 981 the Vikings established a colony on Greenland, which they named for the lush grasses that swept the terrain. Then they sailed even farther west, to Vinland (probably today's Newfoundland), where wild grapes grew in profusion. It is even possible that they traveled as far as the North American continent.

In the thirteenth century, however, the warm weather gave way to a colder climatic pattern. Around 1271, the world's glaciers started to advance. Ice clogged the northern seas, and the Viking explorations ceased.

The northernmost countries of Europe were the first to be affected by this change in climate. Parts of Scandinavia, Scotland, and Iceland were chilled most of the year, and sometimes barely emerged from the snow even in summer.

The dreadful winter of 1315–16 was a bad beginning for what proved to be a disastrous century. The summer was wet and cold. Fields turned into muddy quagmires where no crops could grow. Famine set in. Epidemics soon followed, decimating the European population. This was the first in a series of annual crop failures, famines, and epidemics. In 1348, bubonic plague—the dreaded Black Death—made its appearance in Europe. For the next three hundred years, the scourge coursed its way back and forth across the continent, destroying entire towns, paralyzing cities, throwing desperate peasants into futile revolt. By the time it abated, in the late seventeenth century, an incredible 25 million people had succumbed.

By the 1400s, freezing winters and erratic seasons in Europe had become the climatic norm. During the winter of 1431—the year of Joan of Arc's martyrdom—a mass of freezing air settled above Scandinavia. Bitter blasts of Arctic wind, snow, and ice were directed south over much of the continent. In early 1432, every river in Germany was frozen solid. Glaciers in Scandinavia and the Alps grew menacingly. In France, acres of vineyards were destroyed by frosts. Ice packs cut off the northern seas.

In 1492 Pope Alexander VI asked his bishops why no one had visited the Christian congregation on Greenland for eight decades. The task, if attempted, would have proved futile. The seas were impassable, Greenland was icebound—and in any case its entire population had perished by the 1450s.

In England, the 1560s brought torrential rains that inundated crops. Grapes, which had been cultivated since Roman times—so profusely that English wines were once exported to France—were destroyed by stinging frosts. In 1600, half a million Russian peasants died of starvation and disease when their crops were destroyed.

The glaciers continued to grow at an alarming rate. In the early part of the seventeenth century, entire villages near Chamonix in France were overtaken by ice. In the winter of 1601 and again in 1603, bad weather destroyed Ireland's potato crop, causing widespread famine. Five years later, a handful of English settlers died from hunger and cold when a bitter winter struck the first American colony at Jamestown, Virginia.

The decade from 1643 to 1653 proved the coldest period since the end of the previous great ice age. In 1652, Russian peasants again suffered grievously when bad weather destroyed the grain crops.

During the summers of the 1660s, devastating droughts hit England repeatedly. The Thames dried out to a shallow stream that boats were unable to navigate. London's wooden buildings turned the city into a tinderbox. The consequence was the Great Fire, which destroyed the City and left hundreds of thousands homeless. Happily, however, it drove the plague from London for the first time in three centuries.

Incredibly, the eighteenth century saw the Little Ice Age intensify further. The bitter winters of 1739 and 1740 caused immense suffering in Scotland, Iceland, and New England. In the same years, frost and blight destroyed the Irish potato crop, and thousands starved to death. In 1770, famine and bubonic plague (which continued to harry the Continent), claimed thousands of lives in Russia and Poland.

During the same decade in North America, where the Revolutionary War was being fought, British soldiers dragged cannons across frozen New York Harbor from Manhattan to Staten Island. George Washington and his troops nearly froze to death at Valley Forge.

In 1788, intensive drought in northern France withered the crops early in the summer. On July 13, violent hailstorms pounded the fields, destroying what plants remained. The next year, France suffered an acute grain shortage, which led to the bread riots of 1789, and the storming of the Bastille, and eventually cost Marie Antoinette her head.

In 1800, potato blight, brought on by bad growing conditions, again destroyed Ireland's primary source of food. In 1816, a typhus epidemic wiped out 25 percent of the Irish population. That same year, Yankee farmers watched snow fall on their fields in June, and called it "eighteen-hundred-and-froze-to-death."

Between 1845 and 1847, bad weather once again caused a devastating blight to wipe out the Irish potato crop. Typhus, known as "famine fever," ran rampant. Thousands of Irish people died. Over a million more fled penniless to the United States, in one of the Western world's last great migrations.

After the Irish potato famine, the Little Ice Age abated. Winters shortened. Temperatures started to rise. The seasons became more stable. By the 1860s glaciers had begun to melt. Global weather patterns improved steadily, until, in the past sixty years, they reverted to what we've grown to imagine is "normal."

What Is Normal?

But "normal" weather is in fact nothing of the sort. It is freakishly good. Actual normal weather conditions in the last half million years have been far more like those of the Little Ice Age than the climate man has grown accustomed to during the past ten thousand years. And during this period our civilizations have flowered and grown.

Our tribal drums were pounded, our arrowheads hewn . . .

Our pyramids were built, our Parthenons carved . . .

Our cathedrals reached to the heavens, and our palaces were adorned with Michelangelos . . .

Our volumes of Shakespeare were printed on paper and bound in gilded leather . . .

Our war machines grew hideously efficient, while our children watched television . . .

And these things happened not only because of our genius as human beings, but also because the climatic timing happened to be right. We were lucky. *At no time in the past four hundred years has the climate been quite so warm and favorable as in the past century.*

During this brief hundred years, the surge of human activity has been unprecedented. The population has almost quadrupled. The Industrial Revolution flourished. Diseases have been conquered. Standards of living have skyrocketed. Agriculture, animal husbandry, and fishing have expanded to their absolute limits.

But today our civilization is avidly and thoughtlessly gobbling up the earth's remaining resources. Can the world's population continue to be sustained? Are miracle fertilizers, Green Revolutions, and the latest technology enough?

Perhaps the first step in answering those questions should be to take a close look at the supremely relevant factor the world has lived with from the beginning, but until recently has known very little about: climate.

3

Climate—The Last Great Mystery

The climate of a region on earth is said to be represented by a statistical collection of its weather conditions during a specified time interval. This interval is usually at least two or three decades. . . .

—CIA report

From man's viewpoint earth is the most favored planet in our solar system. It is also the most vulnerable, because it bears life and that life is at the mercy of climatic changes. A sudden frost nips growth in the bud, and the citrus crop fails. Early winters destroy vegetables; unseasonable hailstorms flatten grain crops as a thresher pulverizes husks. Monsoons that fail to arrive cause famine.

Recently the world's climate has been abnormally good. The odds against another fifteen consecutive years' being *as* good as the last fifteen have been put at about one in ten thousand. We've come to depend too much on a benign climate. But what is climate?

The word itself comes from the Greek *clima,* meaning the inclination of the sun's rays. Ancient Greek scientists believed (but could not prove) that the angle at which the sun's energy hits the earth's surface is fundamentally important to our weather patterns. Modern climatologists agree. Today the word "climate" describes weather conditions over a specified time interval, usually twenty or thirty years. "Weather," on the other hand, is the state of the atmosphere—hot or cold, for example, rain or sunshine—at a given time and place.

Records based on accurate observation with instruments go back less than three hundred years. England has the largest continuous record of temperatures, with systematic observations dating back to 1686. The eastern United States has kept records from 1738, and the Netherlands has kept rainfall records since the mid-1730s.

Man is only now coming to grips with the forces that most affect his destiny. Since the late 1960s there have been great advances in analyzing the forces that drive the world's climate. New techniques will increasingly allow us to lift our eyes beyond tomorrow's weather and scan the skies for large-scale changes. It is also possible to look back in time and get a reasonable picture of what the weather was like when dinosaurs roamed the earth, when our early ancestors lived in caves, when the United States was colonized.

Large-scale patterns can be discerned in historical climate, and useful predictions made for the future. But why does the weather work the way it does? What are the real forces behind it? The answers seem certain to remain shrouded in mystery for many years to come.

The new techniques are based on the fact that the earth is a huge heat-exchange engine, a delicately balanced machine of incredible power driven by the energy of the sun. There are five basic components of the machine:

1. The atmosphere
2. The oceans, rivers, and lakes
3. The ice sheets, mountain glaciers, sea ice, and surface snow cover
4. The land masses, including mountains, soil, and rocks

5. Mankind, animals, plants

Each of these components adjusts differently to changing external forces, such as the heat from the sun. For example, the atmosphere does not have a great heat-storing capacity; the sun reaches its highest point in the northern sky toward the end of June—yet the highest temperatures are recorded around the end of July. The atmosphere's adjustment time is about a month. On the other hand, the sea takes much longer to adjust to outside influences. The surface layer—from about thirty to three hundred feet deep—can interact with the atmosphere in a matter of months, but the deep layers of the ocean mix very slowly—it takes about a thousand years for effective heat exchange to take place between the bottom of the ocean and its surface.

At any given moment, the five components mentioned above are changing in relation to each other. This dynamism can be classified as:

1. Heat—the temperature of air, water, land and ice. This is the most important factor in determining climate.

2. Movement of winds and ocean currents

3. Water—moisture, humidity, water levels, clouds, oceans

4. Pressure and density of the atmosphere and the oceans. The Pacific Ocean, the earth's greatest expanse of water, is about a yard lower on its east side than on its west, because of atmospheric pressure.

None of these forces is fixed and constant. They vary, interacting with one another in many ways. Theoretically the variations are limitless. Like the human brain, the earth's weather is infinitely complex.

The climatologists' province is the interaction between the sun, the oceans, the atmosphere, the ice packs, and the winds and the land. Their data comes from the findings of meteorologists, oceanographers, biologists, geologists, and other experts, and from the observations of satellites that constantly scan the earth from their monitoring orbits. Our weather—wind, rain, snow, and hail—is the physical result of the interconnections between these variables of climate. And the climate's priming force is the sun.

The Sun

Solar energy has been calculated at a constant rate equal to 230 trillion horsepower. The grandeur of its vitality is impossible to exaggerate; so is its capacity for good and bad. Another way of describing its effect: In an average eight hours of daylight in New York City, the sun beams 2.8 million watts of energy—the equivalent of 28,000 100-watt light bulbs—onto every acre exposed to it, and that is a *tiny* fraction of the total energy the sun radiates.

The solar energy that is absorbed by the earth's surface is eventually transferred back to the atmosphere; if this did not happen, the surface would warm up and the atmosphere would cool. This vertical heat exchange is vital. Fifty-three percent of the solar energy at the top of the atmosphere reaches the earth's surface as direct and scattered solar radiation (heat and light). Of this, 12–14 percent is reflected back to the atmosphere and space by the earth's surface. However, in high latitudes the polar-ice-covered regions can reflect over 50 percent of incoming heat and light. Of the 47 percent of solar energy not reaching the earth's surface, 24 percent is returned to space by the clouds.

The ability to reflect sunlight instead of absorbing it is called albedo. Thick clouds combined with snow and ice—as you'd find at the poles, for instance—can mirror up to 85 percent of the sun's rays back to space. Even over the equator, thunderstorm clouds more than five miles thick can reflect back as much as 80 percent of the incoming sunlight.

In the transfer of the solar energy absorbed by the earth's surface back to the atmosphere, the chief mechanisms are evaporation of surface moisture (which *takes* heat from the earth) and condensation in the air (which *gives* heat to the atmosphere). Others include eddies of warm air sucked up in the formation of clouds (convection) and air movements caused by turbulence in the atmosphere.

The sun's heat is the trigger that keeps the atmosphere circling around the globe, spreading warm air toward the north and cold air toward the south. Breezes and gales, storms and calms, are all part of this process.

The Winds

The sun pours energy down on the equator. Warm moist air rises from the equatorial waters—the oceans and tropical lakes, rivers, and forests. It sheds most of its moisture, as rain, in the tropical rain belts near the equator before flowing toward the north or south poles. But this mass of warm air gets only about a third of the way to the poles before much of it cools and descends again to create dry belts. (Sinking air does not make rain; as it descends it becomes warm and its capacity to absorb moisture increases.) These are where most of the world's major deserts, such as the Arabian Sahara and the Gobi, lie.

Some of the sinking air fuels the trade winds. These blow toward the equator from the east across the Pacific and Atlantic Oceans, and from north and south of the equator. It was the trade winds that blew Christopher Columbus across the Atlantic to the New World.

The remaining sinking air moves on toward the north and south poles. Wind speed increases. The earth spins like a top, and at the equator its east-to-west spin is more than one thousand miles per hour. In the polar region the spin is much less. As the wind moves poleward it blows more and more strongly from the west. It is the earth's spin that steers the cold and warm air masses of our ever-changing weather system.

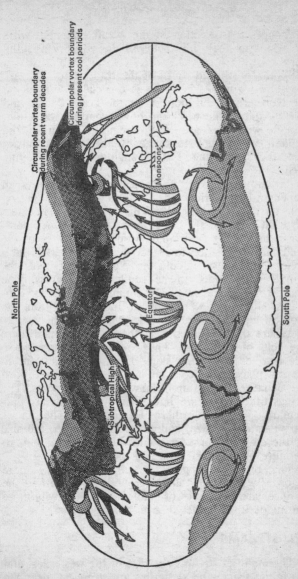

Global weather variations.

It changes the direction of the ocean currents, twists the winds as they flow between the equator and the poles.

To the north and south of the trade-winds belt are the stormy zones. The name is especially deserved in the Northern Hemisphere. Over the north Pacific and north Atlantic the wind can howl from any direction, though westerlies are most common. South of the equator are parallel stormy zones: the Roaring Forties and the Screaming Fifties—nightmares for sailing men.

Air moving toward the poles eventually meets an impenetrable barrier: the huge umbrella of heavy, cold air that caps the poles. This cold-air dome centers on what is called the circumpolar vortex, which comprises the strong west winds around the center of the cold air mass. As winter cools the Northern Hemisphere, this mass of cold air expands and cools the temperate latitudes in its southward progress. The west winds move south. In summer the vortex contracts and allows warmer air to move north from the tropical equator. The west winds move north.

Where the westerlies meet the polar ice mass there are belts of turbulence. They are usually compared to the hem of a swirling skirt. The hem curves out in several places, and cold air is swept in waves toward the equator. Warm air is drawn in to the poles. The waves of cold air are intensified by mountains, particularly in the Northern Hemisphere. The waves vary in number, size, and shape according to the size of the circumpolar vortex and the difference in temperature between the pole and the equator.

The bigger the gap in temperature, the more vigorously the air circulates around the globe and the stronger are the waves of cold air, which is why more storms occur in winter than in other seasons.

JET STREAMS

The main jet streams are part of the westerlies, and are found where the westerlies are strongest, near the polar caps. They can reach two hundred miles per hour. In the Northern Hemisphere the stream passes over North America, Europe, and the U.S.S.R. It circles the world continuously. A balloon suspended in the stream would take about fourteen days to drift around the globe. Lesser easterly jet streams blow at lower levels in the tropics.

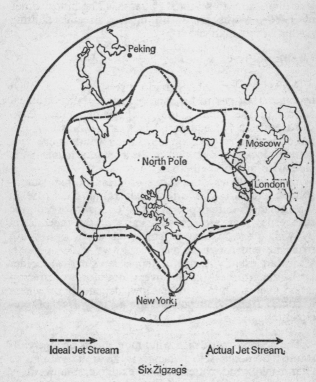

Ideal Jet Stream   Actual Jet Stream

Six Zigzags

(BBC Publications/based on satellite photo taken by NOAA 2.)

As the earth rotates, atmospheric pressure twists and bends the jet stream so that it forms snaking whiplashes that feed warm air to the poles and draw out their cold air.

Natural barriers—the Rockies, Russia's Ural Mountains, and the polar funnel between Greenland and the Scandinavian countries—make the stream zigzag. If it were not for these barriers, its meanders would be even more variable. Such constant natural factors give uniformity to seasonal patterns. (The bitter winter of 1976–77 was the result of the stream's shifting farther south than usual.)

DEPRESSIONS

Areas of low atmospheric pressure, also called cyclones (but not to be confused with violent tropical storms of the same name), depressions move eastward, outside the equatorial belt, sometimes in interlinked groups of three or four. Their winds move in a counterclockwise direction in the northern hemisphere, clockwise in the southern.

With anticyclones (the result of areas of high atmospheric pressure with winds that rotate clockwise in the northern atmosphere, counterclockwise in the southern), they are essential parts of the mechanism by which excess heat is conveyed north and south from the equatorial region.

Warm air usually heralds a depression's arrival, colder air its departure. While it hovers some form of precipitation occurs—rain, hail or snow, depending on where it comes from. Anticyclones are usually free of precipitation.

The general worldwide wind movements change constantly. Sometimes the change is vigorous enough to alter basic wind patterns. For example, the westerly and easterly winds usually predominate, but if the southerly and northerly flows in the waves of the wind (the meridionals) blow more strongly than usual, the force of the winds from the west and east drops sharply. When the meridionals are strong, weather systems move slowly from west to east and form stationary zones of air in the zone between 45° and 50° latitude. These blocks can have a good effect on the weather: In summer, for example, a block over Scandinavia brings warm weather to Europe. But blocks can also cause droughts and heat waves, floods and heavy snowfalls.

The global winds and the way they flow are the mechanisms that bring about climate changes. But they are just mechanisms—not the trigger.

Water, Land, and Ice

THE OCEANS

About 75 percent of the earth's surface is covered by water, most of it in the oceans. These vast reservoirs of energy absorb most of the sun's radiation and play an important role in transferring heat from the tropics to the poles: The warm moist air that rises from their surfaces as a result of evaporation is the start of the world's wind machines. They also absorb heat and retain it for long periods, while underwater currents diffuse the warmth around the world. This is important for the world's weather. For example, the warmth of the Gulf Stream brings a balmy climate to Scotland's west coast. And when the equatorial waters of the Pacific cooled down in the mid-1960s, global winds were weaker the following winter. (Cooler water means that less moisture and heat are available to the atmosphere; in the equatorial zones, it closes the gap between north-south temperatures and the winds decrease.) As a result, Spain and southern California both had lower-than-normal rainfall.

Ocean currents, like the air, swirl and change directions, guided by differences in temperature and saltiness. An example is the cold-water current off Peru. Up until 1970, Peruvian fishermen regularly caught enormous quantities of anchovies in the warm water between the coast and the cold current. Many other

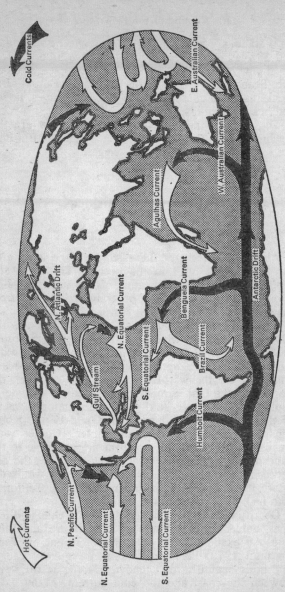

Ocean currents of the world.

countries depended on this catch for their animal feed. By 1972, the fish-processing plants were closed and deserted. Why? The cold current, which normally stayed away from the coastline, had moved inshore. Young fish died in the unfamiliar cold, and the catch that year fell to 4.5 million tons from the normal total of 10 million tons.

Large areas of unusually warm and unusually cold sea-surface temperatures have been noted over the oceans. Dr. Jerome Namias, of the Scripps Institution of Oceanography in San Diego, an ocean expert, believes that these patches strongly affect how the jet stream behaves. Cold water causes air to sink and more air rushes in to fill the vacuum; warm water makes air rise and move outward to cooler areas. This affects the strength and meandering of the jet stream.

The oceans actually lose more water to the skies than they gain from them, providing about 88 percent of the atmosphere's water vapor and receiving only 79 percent of its precipitation. In other words, the oceans give out 9 percent more water than they receive. However, an exact balance is kept by a 9-percent run-off of fresh water from the continental rivers.

LAND

The world's landmass covers only 20 percent of the earth's surface. Deserts have a high albedo, reflecting most of the sun's surface energy back into space. Some experts believe that high albedo reduces the amount of heat absorbed by the desert's surface and causes cooling. However, many desert areas are hot under strong sunlight, because the soil, which has a low heat capacity, warms rapidly even though the albedo is high. The lack of clouds in these regions could well be caused by sinking air in the subtropical high-pressure belts and downwind of mountain ranges —a process that causes air to warm and leads to a reduction of showers and clouds.

Because there are few clouds, the sunlight that reaches the lower atmosphere and the desert surface is strong. A significant amount of direct and reflected solar radiation is absorbed in the deep dust layers that veil desert regions. This warms the air further. And in regions where the desert surface is hot, some of this heat is added to the atmosphere.

The directions in which deserts shift may be determined by where the subtropical high-pressure belts shift, from year to year and over long periods. During our recent cooling these highs in the Northern Hemisphere may well have shifted south and been partly responsible for the increased frequency of drought in fringe zones south of the great desert areas. The Sahel is an example. Hot, wet regions of the earth allow warm air to rise, taking moisture with it and releasing it as rain farther to the north or south. Tropical forests, too, provide moisture, through evaporation, to fuel the world's rain clouds. Difference in temperature between different landmasses sets off a series of chain reactions. Mountain ranges act as barriers to the wind. The turbulence that results when the air rises over them can affect weather in regions a thousand miles away.

ICE AND SNOW

Like deserts, ice packs reflect the sun's light back into space. The temperature of the surrounding air falls as a result, with jet streams and storms tending to form where the large temperature contrasts—between the cold ice surface and warmer regions away from the ice zones—are located. Ice causes change in the wind and alters the ocean beneath it; when ice is melting and when it is forming, it changes the composition of seawater (ice has less salt) and affects the currents. Both snow and ice evaporate during the polar summers, adding to the moisture in the atmosphere.

What Causes an Ice Age?

Scientists understand something of the interplay between the sun, the winds, the oceans, and the landmasses, and how this interchange sparks off changes in the weather. But why an ice age starts is still open to speculation. What could trigger off a slow or quick slide into an ice age? What titanic force has produced the regular hundred-thousand-year Great Ice Age cycle that shrouds great parts of the world in ice?

Only the sun, perhaps, or the earth itself. Or a combination of both.

SUNSPOTS

These immense blotches that move across the face of the sun were first observed in 1611, by the Italian scientist Galileo. Modern scientists have established that these cool areas of the sun—giant magnets created by gigantic storms—vary in size and appear in cycles of approximately eleven years.

Solar flares, jets of white-hot gas, appear when sunspots are at their height and streak hundreds of thousands of miles across space. Sunspots have puzzled scientists for centuries. Some weathermen believe that the Arctic region is slightly more turbulent when sunspots are most active. All climatologists hope that information from satellites will clear up the mystery.

We know that the sun goes through a magnetic cycle every twenty-two years: The magnetic center of the sun reverses its direction during every sunspot cycle and takes twenty-two years to flux back. The sun also undergoes titanic storms, which are signaled by a shorter sunspot cycle. Climatic records reveal another, longer cycle, between peaks of sun-storminess, that lasts about eighty years over and above the shorter cycle.

Another apparent rhythm is a two-hundred-year cycle, which coincides with periods of cool weather.

This finding is based on radio-carbon dating of tree rings; carbon 14 forms when cosmic rays strike organic material on the earth, providing evidence of spells of solar activity. Radio-carbon dating shows that a two-hundred-year cycle apparently reaches back to pre-Christian times. Before about 300 B.C. the rhythm lengthened to a four-hundred-year cycle. This technique shows, and other evidence confirms, that the earth went through cold spells in about A.D. 1300, 1500, and 1700. Now, in the 1900s, we are entering yet another cold spell. Radio-carbon peaks also confirm the dates of the great glacier advances in about 1000 and 3500 B.C.

At first glance it would seem that cool periods on earth and activity on the sun's face coincide. But as yet the time scale is too small, and too recent, for scientists to come to any definite conclusions about a link between sunspots and the earth's ice ages. For one thing, these periods are not all synchronous and widespread. For another, the carbon 14 link appears more tenuous the further back you go in time.

Many experts believe that the "earth wobble" theory put forward by Yugoslav geophysicist Milutin Milankovitch in the 1920s is the most logical explanation yet of the ice-age trigger.

EARTH WOBBLE

The Milankovitch theory posits a continuous, changing relationship between earth and sun. This is how it goes.

The earth's axis in the Northern Hemisphere is tilted toward the sun in July, the height of northern summer. In January, the southern summer, the Southern Hemisphere is tipped toward the sun. At the same time, the orbiting earth moves closer to the sun in January than in July. Thus, with more solar energy available, southern summers are generally hotter than northern ones.

There are also cyclical changes. First, the earth's orbit changes shape every ninety thousand to one

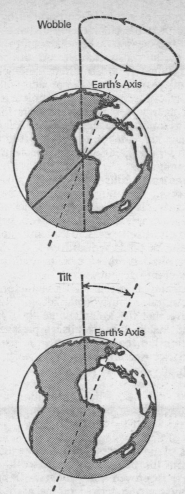

hundred thousand years. From being almost perfectly circular, it slowly changes to become slightly elliptical and then slowly back to circular. The sun's intensity can vary by as much as 30 percent over the cycle—

a change far greater than any produced by the sun itself. Over a year the change is less than 1 percent.

Second, the season when the earth approaches closest to the sun also changes. At the moment this happens in January, the southern summer. In about ten thousand years time this approach will come during the northern summers. A complete cycle (from more solar energy in southern summers through more in northern summers back to more solar energy in southern summers) takes twenty-one thousand years. Right now we are going through the worst phase for northern summers.

Third, the earth rolls from side to side like a ship in heavy seas. The more the earth rolls, the more difference there is between winter and summer temperatures. A complete roll from one side to the other takes forty thousand years. The earth now is rolling upward, and its tilt is reduced. We should be having cooler summers—if other elements, such as the winds, are not taken into account.

The interplay between all these factors, and their variations, is enormously complicated. But many experts believe that the Milankovitch theory is basically correct and that it does establish a pattern in the sun-earth relationship that ties in with the rise and fall of past ice ages. According to the theory, the earth's warm spell may be over.

The Broad Picture

What is climatology's broad picture of the earth's climate from the beginning of our world until now?

Science has proved that about 600 million years ago ice sheets covered most of the earth. Then came a series of long warm periods, when the giant dinosaurs were the kings of creation and the earth's coal and oil deposits were formed. Subtropical flowers grew in Greenland.

About every 250 million years, the northern ice

invaded the south. About 50 million years ago, the earth moved into a faster process of cooling. In the last million years or so, warm periods have occurred only once every hundred thousand years and lasted about ten thousand years.

A CLIMAP (Climate/Long Range Investigation Mapping and Prediction) computer has sketched out a typical average autumn day eighteen thousand years ago. Where New York and London now stand, great ice sheets pressed heavily on the land. Ocean levels were about three hundred feet lower than today. The Gulf Stream had swung south by over a thousand miles.

We are just emerging from a warm spell. What's called the climatic optimum of the earth's present interglacial period started about eight thousand years ago. By about 4000 B.C., the average temperature in the northern hemisphere was about 1.8–3.8°F higher than it is today.

Just why the climate should swing from one extreme to another is still a mystery. There are almost as many theories about climatic mechanisms as there are weathermen. Some experts think the climate will change of its own accord to control its systems and interactions—as a thoughtful computer will automatically adjust incorrect data. Others believe that the trigger is related to outside phenomena, like sunspots or earth wobble.

Still others are skeptical of ever finding the answers, and add that perhaps we do not yet know if the correct questions are being asked.

Professor Edward N. Lorenz, of the Massachusetts Institute of Technology, once summed up the problems that climatologists face: "Can the flap of a butterfly's wings in Peru cause a tornado in Iowa?" He also had a less rhetorical point: that a casually dropped smoldering cigarette butt could start a brush fire—which would make local thunderstorms increase in intensity. The climate's trigger—including the one that sets off an ice age—could be just that sensitive.

4

The Upside-Down Greenhouse

Why has the earth cooled? There are three main factors involved affecting how much sunlight reaches the earth and how much is re-radiated into space: volcanic dust, man-made dust, and carbon dioxide.

—CIA report

Experts argue endlessly about exactly what effect man will have on the earth's climate. Yet, as we have already seen, one fact is certain: From the start of history, man's treatment of his environment has influenced weather.

Thousands of years ago, the Sahara desert was lush and fertile. Overgrazing of its pasture lands and slash-and-burn farming helped turn it into the world's largest sand bowl. The result? Warm air over the desert reduces clouds and showers and keeps the desert dry.

Cities make their own weather. If you work in downtown Washington or New York, you'll have 16 percent more rain than your family out in the suburbs —and you'll suffer through more summer thunderstorms. La Porte, Indiana, downwind from the industrial Gary-Chicago complex, gets an incredible 246 percent more rain than Gary-Chicago as a result of this phenomenon.

Climatologists tend to agree that people are "creating" weather. Where they disagree is over the key question: Is the earth heading toward an icy future or a fiery one?

The "cool-earth" men point to a curious temperature pattern over the past seventy or so years, the tail end of the world's abnormal weather. In the U.S. average annual temperatures increased by 3.5°F between 1920 and 1954. If the rise was caused by increased radiation from the sun, temperatures should have soared all over the world. But they did not. In areas such as the Hudson Bay region, northeastern Australia, Indonesia, parts of central Asia, and central and southern South America, the mercury dropped. Thinly populated, and far from any kind of industrial development, these regions are unaffected by man's activities. Their cooling gives strength to the theory that man is helping to heat the earth's atmosphere.

The "hot-earth" men believe that man's pollution of the atmosphere is helping along a process that could end with the north and south poles as hot as the tropics are today, and with the tropics so hot they would be able to support only cold-blooded lizards and insects.

Given the possibility of a coming ice age, pollution could be our short-term salvation.

Some of the ways in which man generates heat and warms the atmosphere are obvious. Anyone who lives or works in a city knows that it's warmer there during a cold winter than in the suburbs or out in the country. Concrete and brick buildings and asphalt roads act as giant storage heaters. They absorb the sun's heat and release it slowly. Waste heat—from industry and central heating, power stations, and auto emissions—adds to the dome of warm air that hovers over the world's big cities.

In densely populated areas and industrial complexes the amount of energy released is about 10 percent of the heat supplied by the sun. New York City gives out seven times more heat than it receives from the sun's rays. England's infrequent snow usually melts when it reaches the outskirts of sprawling London, with its population of 7.5 million.

The Invisible Greenhouse

Strangely enough, the earth's most effective and crucial warming mechanism cannot be seen, touched, or even smelled—although it is present everywhere and all the time. It is an invisible gas that makes up an infinitesimal proportion of the gases that form the atmosphere: carbon dioxide.

Carbon dioxide acts as a glass greenhouse. It allows the sun's short-wave, or visible, radiation to pass through to the earth's surface, then absorbs the longer-wavelength infrared rays emitted by the earth's surface and gases. In other words, it creates a greenhouse effect on earth by trapping the warmth of the sun. But, unlike a real greenhouse, carbon dioxide is not just a civilized luxury. Without it (and water vapor, which has the same effect) the earth would be at least 90°F colder than it is today—a frozen planet without life.

Carbon dioxide is an important product in the decaying of organic materials, such as are found in bogs and marshes, and is released to the atmosphere when these dry up. About 366 billion tons of the gas would be released if all bogs and marshes vanished. Weathering of rocks is another source of the gas. Rocks contain a million times as much carbon dioxide as man has released so far.

In order to grow, plants inhale carbon dioxide by day and exhale it by night. Scientists at the Mauna Loa Observatory in Hawaii have recorded a decrease in the earth's carbon dioxide level during the northern summers, when growing plants absorb the gas, and an increase in winter, when plant activity diminishes. Plant breathing does not alter the carbon dioxide content of the atmosphere over long periods of time. However, plants, especially trees, are reservoirs of the gas. And when trees die and rot, their carbon dioxide returns to the atmosphere. It is part of a delicate process that ensures a constant supply of this vital gas in the air.

The level of carbon dioxide in the atmosphere was relatively stable until just about one hundred years ago, when man began making use of the earth's fossil fuels—coal and oil. When these are burned, they release heat—and carbon dioxide. In just over a century, our furnaces and coal fires, aircraft and cars, industrial plants and ships, have added 360 billion tons of carbon dioxide to our atmosphere. That is an increase of about 10 percent since the last century. Experts believe that the next 10 percent increase will take only twenty years, and the one after that only about ten years. At that rate, the atmosphere's carbon dioxide could be doubled in just fifty years. And once carbon dioxide is in the air, it is there to stay. Science does not know how to clean the atmosphere. Even in ideal, unpolluted conditions, the carbon dioxide cycle from air to earth and back to air takes many centuries. We have added to the level of carbon dioxide and started a process that will take a thousand years to right itself. And that's if we stop burning coal and oil right now, today.

The more carbon dioxide there is in the atmosphere, the hotter our greenhouse will become. Much of the carbon dioxide added to the atmosphere is absorbed by the oceans, which are the earth's greatest reservoir of the gas. However, there is a limit to the amount the oceans can take up, and we could well be approaching that limit. If we do reach it, excess gas will stay in the atmosphere instead of going into the oceans—and the carbon dioxide increase will be exacerbated.

In addition, as the planet warms more carbon dioxide will be released by the ocean to the atmosphere—just like soda pop warming up. Some scientists believe that great waves of carbon dioxide have

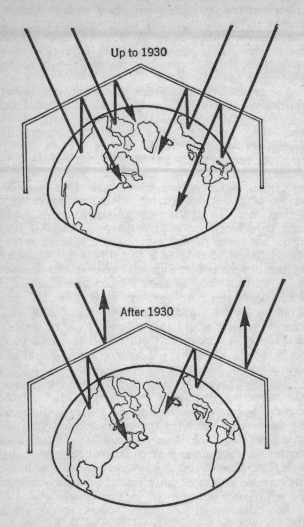

The Greenhouse theory.

waxed and waned in this manner on Mars—in amounts greater than our *total* atmosphere.

Right now, earth could be heading for a 4.5°F rise in just fifty years. Even a 1.8°F rise around the globe would represent a climate change as dramatic as any experienced in the last thousand years.

The hot-earth men believe that carbon dioxide is helping us along to a "Venus situation," so called because that planet's 900°F temperature is the result of the great amount of carbon dioxide in its atmosphere.

The cool-earth men say, "Great. A small rise in temperature is just what we need to offset earth's present cooling trend." But they do modify their optimism. In global terms, man contributes a tiny amount of heat to the atmosphere. Admittedly, his contribution affects local and even regional weather patterns. But the sun gives out six thousand times more energy than all man's power stations, furnaces, engines, and industrial complexes put together. One thunderstorm releases more energy than New York City gives out in an entire year.

Hot-earth men and cool-earth men agree that the rise in carbon dioxide will be a vital factor in determining the earth's future temperature.

Climatologist Reid A. Bryson believes that its rise has, in the past, helped balance out the cooling caused by two other main factors that affect the earth's temperature: dust from volcanos erupting, and dust created by man. "According to his theory," one CIA report says, "the earth would have cooled due to this dust even more than it has if it had not been for measurable and increasing amounts of carbon dioxide which man has put into the atmosphere by burning fuel (the greenhouse effect)." (See Appendix 2.)

The Dust Veil

The strength of the sun's rays when they reach the earth depends on the clearness, or transparency, of the air they pass through. At least one third of this energy is reflected right back into space—and that proportion rises to 85 percent if the rays strike ice and snow. But unless you're on a polar ice cap, or skiing in Denver, you can be pretty sure that, the stronger and brighter the sunlight, the hotter the weather will be. Clouds or particles of dust screen off the heat and the temperature drops.

Scientists use the word "aerosol" to describe particles suspended in a gas. These invisible fragments of matter can be tiny droplets like the water in clouds or fog. They can be solid: ice, sand, dust, soot. Or they can come from chemicals sprayed out of aerosol cans. Like carbon dioxide, they make up only a minute proportion of our atmosphere. But, natural or man-made, they affect the strength of the sun's rays on earth.

VOLCANIC DUST

Volcanos are the world's natural dust-makers. Scientists know that the dust spewed out by volcanic eruptions lowers temperatures on earth. The amount of sunlight reaching large parts of the world can fall by 2–3 percent a few months after a large explosion, and it can be a year or more before the air is clear again. The percentage may not sound like much, but it can have a major effect.

When Tambora, on the island of Sumbawa in the East Indies, exploded in 1815, average temperatures in the Alps dropped by 2°F. Mount Agung on Bali erupted in March, 1963; in the upper atmosphere its dust absorbed the sunlight to increase the temperature by 5°F. At the same time, it filtered the sun's rays to cool the earth's surface. In 1964 and 1965, sur-

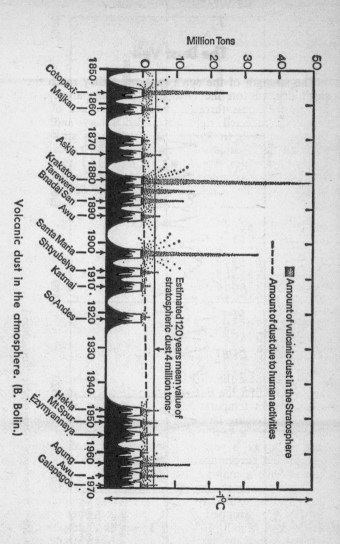

Volcanic dust in the atmosphere. (B. Bolin.)

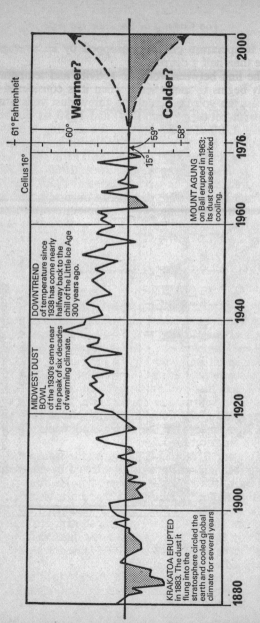

KRAKATOA ERUPTED in 1883. The dust it flung into the stratosphere circled the earth and cooled global climate for several years

MIDWEST DUST BOWL of the 1930's came near the peak of six decades of warming climate.

DOWNTREND of temperature since 1938 has come nearly halfway back to the chill of the Little Ice Age 300 years ago.

MOUNT AGUNG on Bali erupted in 1963; its dust caused marked cooling.

Warmer?

Colder?

(redrawn from a National Geographic illustration.)

face temperature on earth dropped by an average of more than half a degree.

The link between volcanic activity and temperature really begins to make sense when it is considered that at least ten volcanic eruptions threw dust veils around the earth between about 1750 and 1850, at the height of the Little Ice Age. And the cooling trend that started in the 1940s coincided with an increase in volcanic activity in the 1950's. There were twenty great volcanic eruptions between 1950 and 1970, compared with just seven between 1920 and 1940 over the Northern Hemisphere. Conversely, the warm spell at the beginning of the twentieth century was an era when volcanic activity waned, and carbon dioxide increased.

MAN-MADE DUST

All over the world man is producing dust and smoke —from his industrial plants, from farming methods that suck the land literally dry and release dust to clog the atmosphere, from wastes and natural gas burned at oil wells. Soot from the burning of coal and oil is scattered into the air. Forest fires, power stations, and the overgrazing of arid land all help to build up a veil of dust in the skies. And, like natural volcanic dust, the human version screens the sunlight from the surface of the earth and cools down the earth's temperature.

Professor Bryson estimates that the transparency of the atmosphere was not much affected by man's activities until 1930. (See Appendix 2.) He first realized the importance of man-made dust on a visit to India. Flying over the country, he was fascinated by a blue haze that shrouded the entire continent, stretching up to about twenty thousand feet. At first visibility was about seven miles, but when he flew down to Saigon it declined to only a mile and a half.

For once, industry was not entirely to blame. Similar blue hazes veil Brazil and Central America, and are taken for granted by meteorologists. These are usually between about three thousand and nine thousand feet above the ground, and visibility seldom reaches more than three miles. These hazes are caused mainly by desert dust and farming methods that create dust by denuding the soil of the protective vegetation.

They are natural high-altitude equivalents of the smogs that cover the world's large cities. Los Angeles and Tokyo are notorious for their photochemical fogs, but they develop near any big city: New York, Chicago, Rio de Janeiro, and Sydney. Experts are divided as to whether smog ultimately warms or cools the air.

NATURE'S DUST

When ocean waves break, bubbles burst in the sea, and small water droplets ejected into the air evaporate and turn into invisible particles of salt. These aerosols are found mainly on coasts and over the oceans, where, put together, they would add up to 10,000 million tons of salt.

On land, dust is formed by wind blowing tiny fragments of soil, especially minerals and quartz, into the air. Plants exhale gases that may later be transformed into solid particles. Even clear, high-altitude mountain air contains aerosols, although in infinitesimal amounts.

EFFECTS OF DUST

Aerosols affect the earth's climate by absorbing some of the light from the sun and scattering it back to space. The more aerosols—or dust particles—in the air, the less sunlight we receive.

U.S. and Swedish scientists have observed a persistent haze covering the eastern United States and Europe—an area that makes up 1 percent of the earth's surface. They say that the effect on sunlight is "not small." The haze could shorten the growing season in the upper Mississippi basin by a week, perhaps more. Robert J. Charlson, Professor of Atmospheric Chemistry at the University of Washington, concludes that we could be heading for a temperature drop of about 0.9°F on the evidence just of these hazes.

Every year 296 million tons of man-made aerosols are produced on just about one fiftieth of the earth's surface. That's nearly 84 percent of all the particles thrown into the air all over the world. About 80 percent of soot, dust, smoke, and chemical particles remains in the air for several days.

Observatories throughout the Northern Hemisphere report increases in atmospheric dust—and a reduction in the amount of sunlight that reaches the ground. In the U.S.S.R. 10 percent less sunshine reached the earth in 1967 than in 1940.

CLOUDS

Weathermen have calculated that a 1 percent increase in the world's cloud cover could drop temperatures by 1.4°F—four times the total drop of the past twenty-five years.

Man is busily helping to increase the world's cloud cover. According to one theory, jet planes have already affected the ocean beneath their routes: Their vapor trails build up clouds; the clouds reduce the amount of sunlight and heat; the ocean grows colder. This cloud growth is not just a trick of the Atlantic atmosphere. The National Climate and Atmospheric Research Group at Boulder, Colorado, has confirmed that wherever jet aircraft trails form they cause high-level clouds over large areas.

Today, low clouds cover about 31 percent of the earth's surface. If this percentage rose to just 36 percent, average surface temperatures would drop 7.2° F. We would be in an ice age. Clouds form on aerosols; man-made particles could help to make more clouds.

Then, too, scientists have worked out that at our present rate of pollution we could increase the dust content of the air by 400 percent within a hundred years. That would cause a reduction of sunlight that would, in turn, cause the earth's temperature to drop by 6.3°F. Again, this alone might cause an ice age.

A good part of the dust increase is caused by man. And, unlike the carbon dioxide added to our air, aerosols disappear in only a few weeks. We could, if we wanted to, control if not remove this form of pollution.

5

The Handwriting on the Wall— The Unsettled Forecast

Perhaps the simplest worry is marked variation within the prevailing weather patterns . . . in periods when climate change is underway, violent weather—unseasonal frosts, warm spells, large storms, floods, etc.— is thought to be more common.

—CIA report

We are in for unstable weather. And unstable weather means that all parts of the globe will be hit by extremes of climate. Over the next several decades droughts and floods, hurricanes and tornados, killing frosts and throttling snows are certain. They will occur far more often and far more randomly than ever before.

The consequences for man will be dramatic, even drastic. Famine will kill many millions along the monsoon belt of the world as the monsoons repeatedly fail. In the sub-Saharan desert zone, the Sahelian region, 24 million Africans have already been badly hit by the inexorable advance southward of the Sahara Desert. Three million have starved to death since 1970, when the sand began its one-mile-a-year slide toward the equator.

Its arid breath has already scorched to death once-fertile grazing land, and driven the nomadic Tuaregs into the hungry, overcrowded towns. Mass migration to the notorious Black Volta valley in West Africa has taken place. There, river blindness, carried by swarms of blackfly that breed in the rivers, is rampant. At least one out of every four adult males is affected.

About 10 million Africans are on the move. If the Sahara continues its advance, Mali, Upper Volta, Niger, and Chad could all become uninhabitable within a few years. Senegal and Mauretania would be reduced to mere coastal strips.

In other parts of the world, floods will obliterate crops and make vast tracts of growing land unusable. Longer and colder winters, short growing summers, and early and late frosts will reduce the growing season by weeks. The new crop strains of the Green Revolution will be badly hit. They are dependent on the abnormally warm weather of the 1950s and 1960s, and offer less resistance than older, indigenous strains. Dr. J. McQuigg, of Missouri University, believes they cannot survive outside the narrow spectrum of our "typical" temperatures and rainfall. Severe world food shortages are inevitable within ten to fifteen years.

To add to the urgency of the problem, the world is running out of farm and grazing land. India, China, Southeast Asia, and Japan have practically no land reserves. Most of China's growing area is already double- and triple-cropped. Every year the world loses several million acres of arable land through soil erosion and urban spread. This loss can only increase in times of increased drought.

To make more land available will require massive projects of river diversion, swamp drainage, dams, and irrigation canals. But the necessarily huge capital outlay could well be wasted if the weather takes another random turn. With unstable weather, economic planning gets lost in a scramble to feed people now—whatever the cost.

The Wisconsin Study makes the dilemma of the future painfully clear: While underdeveloped countries starved, the richer countries, including Canada,

Britain, Italy, France, Germany, and Holland, would be forced to concentrate their resources on their own people.

Although world reserve grain stocks are building up, a long period of exceptionally fine weather will be needed to bring them back to the 1972 level of a sixty-nine-day supply. The chances are we won't get it. Although some weathermen believe the world is warming, not cooling, they all agree on one point: The next few decades will be a period of fluctuation. Then, too, warming, by increasing the amount of moisture in the atmosphere, would melt ice caps and lead to increased precipitation.

Reading the Weather

Early man tried to influence weather by sacrificing virgins to the tribal gods and by staging rain dances. North American Indians smoked special pipes and shot arrows at the clouds. They prayed, danced, and chanted. The Choctaws hung a fish around a tribesman's neck, then stood him in the nearest stream until the rains came. The early Greeks believed their gods changed the weather at whim, that Zeus hurled his thunderbolts willfully.

Over the centuries, farmers and sailors learned to read the signs of change. Dandelions and other flowers closed their petals whenever a storm approached. The movement of clouds over the seas and the behavior of sea birds alerted mariners to squalls or dog days. Any country boy could tell the temperature (or thought he could) by counting how many times a cricket chirped in fourteen seconds, then adding 40.

Today, short-term one-to-three-day forecasting is reasonably accurate on the landmasses of the Northern Hemisphere. For instance, American forecasters can predict when a cold air mass coming down from Canada at, say, five hundred miles a day will meet warm air and produce rain. But in coastal areas short-term prediction is more complicated. There are more factors to consider, and they are more variable. In Britain, for example, temperatures can go up and down like a yo-yo day after day.

Beyond three days the forecasters themselves admit they are in the land of speculation. A thirty-day forecast is better than a random guess—but not by much. There's an obvious need for longer-period forecasts, say, one to five years.

Climatologists, like all weathermen, depend on statistics and on highly sophisticated tools:

Numerical models. Computers and mathematical models simulate the weather and analyze the synthetic data that result. Today's computers are comparatively slow for long-term climate experiments; speed and size still need to be improved. More important, the models and data that go into them must also improve if climate-scale predictions are to be possible. But they have led to new discoveries—for example, that a decrease in the northern ice might make Europe cooler.

Analysis of past climates. Some weathermen believe this is the most fertile approach. This kind of analysis reveals past weather patterns and is useful as a basis for long-term climatic speculations.

Satellites. Scores of American, Russian, and European weather satellites are aloft at this very moment. They provide global coverage every day. Their chief value is that their observations supply weathermen with a continuous record of current weather patterns and changes over areas not otherwise accessible, such as the southern Pacific Ocean. Some of their recent triumphs include measuring the real extent of the polar ice fields and the heat balance of the globe. Climatologists look to satellites for important breakthroughs in their knowledge of the weather's most important constituent: solar energy. At present, we know the sun's energy output to an accuracy of 1 percent. We need to know it to an accuracy of 0.1 percent.

Rapid digital computers. With vast memory banks to store and process the inflowing data, they are helpful. But there are large gaps in our knowledge, especially of the Southern Hemisphere. Even the computers of the next two decades will not be able to unravel some aspects of the weather mystery.

This is due partly to lack of data, partly to the shortcomings in the physical models that use the data, and partly because the complex calculations necessary for an accurate computerized weather forecast take more time to process than the weather takes to happen.

And we need to know. You hardly notice a 2°F deviation of temperature in a month, but a farmer in the U.S. spring-wheat zone does. The margin for his crop's pollination is so fine that a summer hotter by just 2°F will cause losses of over a hundred million dollars.

The Danish government invested heavily in Greenland's fishing industry after learning that huge schools of codfish had moved north into Greenland's waters. Only after making the investment did it discover that the water had turned cooler—and the cod had turned back.

Hot or Cold?

WARMING UP

Many hot-earth men believe that global temperatures will rise by at least 3.8°F by 2020, given that the volume of carbon dioxide is doubled in the next fifty years. If this happened, ships could well sail the entire Arctic Circle, and the melting of the polar ice caps could cause the sea level to rise by two hundred to four hundred feet. London and New York would vanish. So would Rome, Paris, Brussels, Antwerp, Marseille, and hundreds of other cities. Trees would grow in Alaska and Siberia; cattle would be raised on what was once tundra.

As already described, increasing carbon dioxide in the atmosphere is an important factor in raising the

71

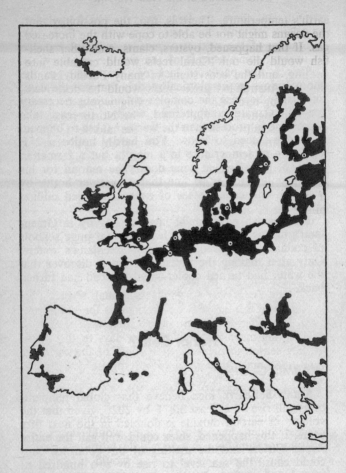

■ Flooded areas
◎ Cities with more than 1 million inhabitants

Areas of Europe that would be affected if the polar ice caps melted. (B. Bolin.)

earth's temperature. There is, too, the possibility that the oceans might not be able to cope with the increased gas. If that happened, oysters, clams, and other shell-fish would die out. Coral reefs would crumble into nothing, and the ecosystems of many Pacific Islands and the Australian barrier reefs would be destroyed.

The long-term effects of warming would be more rain and higher temperatures in the Northern Hemisphere, a semitropical climate for the middle latitudes. The countries close to the equator would become even warmer, with even higher humidity.

Weeds flourish in heavy rain. They can wipe out as much as 40 percent of crops, despite the efforts of technology—and plain old weeding by several million pairs of hands.

When it's wet, birds and rodents go far afield in their search for food, ravaging crops and making crop storage very difficult.

Finally, rain excites disease, fungi, and bacteria to redouble their attacks on crops.

William A. Kellogg, of the National Center for Atmospheric Research at Boulder, takes the optimistic view. He believes that a global increase of 1.8°F would bring more rain and add ten days to the crop-growing season in the temperate zones of the earth. It would also make arid wastes fertile again.

COOLING DOWN

In a cooling world the Northern Hemisphere would become colder. On average, there would be cooler and shorter summers in the middle latitudes, colder and longer winters. A cold north would mean drier tropics and severe droughts in many parts of the world. As the ice sheets expanded, freshwater resources would drop. Spring melting would cause floods. The vital monsoons would become weaker and arrive less often, with unpredictable irregularity.

The CIA believes that perhaps our gravest short-term concern should be, as it has been recently, marked variations in prevailing weather patterns—one of the side effects of a cooling trend. They conclude:

"The US middle West has had moderate to severe droughts every 20–25 years—e.g. 1930s, mid 1950s—as far back as the weather records go. If this pattern holds, the main US granary (now also the mainstay of the world grain trade) could expect drought and consequent crop shortfalls within the next several years." (See Appendix 2.)

In America's Midwest there could be a return of the disastrous dust bowl of the 1930s. There will surely be prolonged drought. Most hot-earth men stress floods and low crop yields throughout the western world.

England's Professor Hubert Lamb notes the renewed increase since 1961 of the Arctic sea ice. This has already caused disruption of the northern sea routes in Soviet and Canadian Arctic waters; a substantial rise in the level of America's Great Lakes and the lakes of equatorial eastern Africa since 1961; increased drought in the Sahel region, and reduced rainfall in the densely populated northern countries lying in the zone between 45° and 75°N–latitudes that cover the U.S.S.R., most of Europe, and Canada.

Could a change in dietary habits defeat the famine? Theoretically, yes. The world could learn to shuffle its food around—why shouldn't the southern Chinese eat something other than rice when the monsoons have failed them? Or the Indians take to grain? But a common-sense idea does not easily overcome ingrained habits and taboos. In India in the mid-sixties many thousands of people in the State of Kerala starved to death from lack of rice—despite the fact that there was plenty of wheat in its granaries. This is not the place to offer detailed solutions to the problem. But there are certain starting points to consider. Could people be reeducated to eat "new" foods, for example? Or would an international food bank with powers to switch acceptable foodstuffs between famine areas be the answer?

Making Rain

Artificial rainmaking began in 1946, when General Electric researcher Vincent Schaefer exhaled into a freezer and watched a miniature cloud form. He added frozen carbon dioxide (dry ice) and watched excitedly as snowflakes fell in his Lilliputian weather simulator. When he experimented at higher temperatures, his tiny cloud produced rain.

Within six years, clouds over a quarter of America had been seeded with dry ice or with its alternative, silver iodide. In 1957 the President's Advisory Committee on Climate Control found that seeding could boost rainfall by up to 17 percent.

Certainly, it has been proved that seeding can play a significant part in influencing the weather. For example, the Bureau of Reclamation at Colorado State University, as well as other institutes, has found that a carefully controlled cloud-seeding program increased the snowfall in the Rockies by more than 15 percent.

During the Vietnam War the U.S. Department of Defense spent about $40 million on a seven-year seeding program, in an attempt to waterlog the Ho Chi Minh trail, the main supply channel from Hanoi. Although the Defense Department claimed they caused a seven-inch rainfall in one area during June, 1971, it, like many other efforts in Vietnam, did not produce the desired results.

There are other rainmaking methods. The Russians and Chinese have bombarded clouds with artillery shells and rockets to coerce them into yielding rain —with confusing results. One report says that, unexpectedly, the bombardment greatly reduced the ferocity of a hailstorm. Unfortunately, hail damage to crops in another area, near the Bulgarian border, increased nine-fold. It may have been coincidence, maybe not. It's impossible to prove cause and effect.

But the revolutionary discovery of seeding is not

the answer to drought. Clouds first have to be present before they can be seeded.

Covering the desert with blacktop or some similar substance is another suggestion. The land would then absorb more of the sun's rays than it does at present, the theory goes, which would cause more evaporation, and eventually rain.

Weather control or modification has its critics. They argue, for example, that seeding tropical clouds could disrupt the weather by dispersing equatorial heat. If typhoons and hurricanes are interfered with, destructive as they are, their major role as heat exchanges between the tropics and the middle latitudes is disrupted.

Unplanned Control

Man is a puny puppet of the all-powerful weather machine. It is possible that one day we may be able to control small, local weather variations for our benefit. It is equally possible that our pollution of the atmosphere could produce overwhelming disaster for the human race.

But there is another of man's effects to consider: the "puncturing" of the earth's ozone layer. Ozone is not just the "smell of the sea." It is an odorless, colorless gas that forms a screen around the earth, about twenty miles up, and diffuses and modifies the sun's blistering ultraviolet rays. Without its protection, all human life would cease.

Obviously, tampering with the ozone layer is dangerous. Yet that's exactly what we have been doing. Our seemingly innocent tools are everyday articles: hair sprays, deodorants, air fresheners—all substances forced out of their canisters by air propellants, which scientists call fluorocarbons. Many experts believe that fluorocarbons in some way puncture the ozone layer and let in thin beams of ultraviolet rays. One estimate is that the use of these gases is growing by more than 20 percent a year. For each 1 percent decrease in the ozone layer, we can expect the incidence of skin cancer to rise by 2 percent. Other evil effects include premature aging, eye damage, and an increase in cases of severe sunburn.

Engineering projects are another danger. Dams and irrigation canals have brought fertility to arid regions and, if well planned, do not usually cause serious ecological damage. But larger-scale projects, such as the diversion of rivers, are inherently much more dangerous. They interfere with the grander designs of climate.

Soviet climatologists effectively stopped a plan to divert several of the great Siberian rivers to feed irrigation schemes so comprehensive that they would have increased the U.S.S.R.'s grain yield by a third. These rivers empty into the Arctic Ocean, where the lighter fresh water spreads out over the salt seawater and allows the Arctic seas to freeze over. Less fresh water would mean less sea ice. The result could be less rainfall over continental areas in the middle latitudes. The Soviet weathermen convinced their government of the very real perils of the plan, but it remains to be seen whether an intolerably prolonged drought in Siberia—the kind that cooling would bring—would force the U.S.S.R. to go ahead with the project.

Over the centuries man's farming habits have gradually produced distinct regional changes: the reduction of parts of South Africa and Southwest Africa to semi-desert; the leveling of the mountain forests in many countries to make room for arable farming (the Savannah grasslands of the tropics are nearly all man-made).

And, of course, as the world's population increased, man continued to affect his environment—and his climate. In 1971, thirty scientists from fourteen countries met in Stockholm to work out as best they could how man influences his climate. Their conclusion, subsequently backed by a great deal of evidence, was disturbing.

"There can be little doubt," their report states, "that

Man, in the process of reshaping his environment in many ways, has changed the climate of large regions of the earth, and he has probably had some influence on global climate as well, exactly how much influence we do not know."

Whether our future is cold or benign—and the evidence points to a cooling—it is essential that we find out about it quickly.

6

The Unbalanced Checkbook—
People, Food and Weather

Man's old age concern about adequacy of food supplies
has resumed with particular urgency since the crop
failure of 1972.

—CIA report

In 1972, India's monsoon was poor. China had a
drought in the north and floods in the south. The
U.S.S.R. had both a drought and an exceptionally
short growing season. Parts of Central America and
Africa also experienced severe drought. And parts of
the world were starving.

U.S. grain stocks were bulging. We depleted them
enormously. In retrospect, to allow reserves to get so
low seems naïve. But there was money to be made,
and the grain brokers were keen to make it. The result
was that America was left with such low stocks that in
subsequent years we have not been able to make up
for crop failure in any major area, even when we put
the entire United States land reserve back into production.

By early January, the United States Government
Crop Bulletin was predicting that 1977 might well
be the first year in which the world has had a grain
surplus in five years. But the severity of America's

winter combined with the great Western drought and the worst dust storm in twenty years could well wipe out this potential surplus.

Scientists are in agreement that the United States now faces the most severe drought conditions of any place on the globe. In such circumstances we simply may not have enough food to feed the world's hungry billions.

Why, in our technological age, can't something be done about it? A great deal has been done, but still our increase in food yield is not enough to overcome one appalling fact: The world's population is growing faster than our ability to feed it. The world's population was about 2 billion in 1930. Today it is 4 billion. It will be 6.5–8 billion by the turn of the century. This increase is not just because birth rates are high in many areas of the world (although in most modern industrial societies the population explosion is a thing of the past; America's population is growing older, not growing quickly, and in some countries such as West Germany population is actually declining). The increase is due to the rapid decline in the death rate, particularly among infants.

Five hundred million people in the world today suffer from some form of hunger. In sub-Saharan Africa, Brazil, India, and Bangladesh they are dying. The nutritional situation is only slightly less serious in Honduras, Burma, Burundi, Rwanda, the Sudan, and Yemen. Furthermore, poor harvests caused by poor weather have now caused serious food-supply problems in Nepal, Somalia, Tanzania, Zambia, the Philippines, and Mexico.

Like the rich getting richer and the poor getting poorer, the populations of the less-developed countries, including China, are growing faster. Right now they account for about 70 percent of the world's population. By the year 2000 they will account for 80 percent.

"Food" really means "cereal." Cereals are the staff of life. They provide almost 50 percent of the protein in the world's diet. When the well-fed world sits down to its breakfast each morning it thinks of cereal as Kellogg's and Post Toasties. But cereal means all kinds of grain: wheat, maize, rice, barley, oats, rye, corn, and so on.

Half of all the cereal America produces goes to feed cattle. To produce one pound of meat takes sixteen pounds of grain. So the meat eaters of the world consume their food twice—first in the form of grain to feed the animal, and then by eating the animal itself. Each North American consumes over one ton of grain equivalents each year, only 70 percent of which is eaten directly. And, on average, every man, woman, and child in the U.S. eats over one hundred pounds of meat each year. In dramatic contrast, the average citizen of the People's Republic of China eats the equivalent of only four hundred pounds of rice a year —over two thirds of which is eaten directly.

Walter Orr Roberts, former President and Director of the National Center for Atmospheric Research, told a 1976 congressional subcommittee on environment and atmosphere that he had retired early to concentrate on the problems of world food. He had done this because the history of past climates had led him to believe that we could not expect as many favorable growing seasons in the future as we have had since 1960. "Moreover," he adds, "there are over 70 million more mouths to feed this year. Shortly after the year 2000 there will be two people on earth for every one here today . . . I maintain that housing and feeding this world, and the second world quietly arriving, is the greatest challenge to human ingenuity that we face for the balance of this century. To me, sustained droughts like the drought in the Soviet Union in 1972 are the worst natural scourges known to humanity. They have a gross impact on agriculture, they can trigger land erosion and water supply depletion, and they are the cause of vast human misery. One only needs to look at 1972 and the 'domino effect' of that to be aware of it."

What happens in the domino chain of a severe drought? While Americans in the Northeast were freezing, people in the Southwest and the West were suffering drought. The dry winter of 1976–77 in California was the second in a row. They are only just waking up to its effects. On February 21 eleven governors and representatives from the six western states, including Oregon and Washington, met to establish a task force aimed at speeding up federal aid. The water shortage was, they announced, "real, immediate, and could be devastating." They wanted $525 million for new dams, canals, and aqueducts.

Drastic reduction in irrigation water has already cost California cattlemen $460 million. California supplies 40 percent of all our food. While Florida lost only 30 million boxes of damaged or frostbitten oranges, California may lose up to half of its $9 billion agricultural industry. The state needs about 32 million acre-feet of water for irrigation. This year, if it is lucky, it will get half of that—and that half depends on "normal" rainfall between January and March, the wet season.

At the same time that Carter's new administration is trying desperately to create jobs, California is starting to lay people off. For every farm worker laid off, three more employees in related industries are affected. Take Del Monte alone—consider the tin for its cans, the cartons for packing, the trucks for distributing . . . another domino chain begins.

The government admits it does not know what to do about it—except try to live on less water. It can't even drop ice on clouds to create rain. There aren't any clouds.

In February, 1977, one farmer told the *New York Times:* "This thing gets scarier and scarier. One thing being learned in this drought is that we are our brothers' keepers now, that no one is going to come out whole from the drought. I can't bring myself to believe the good Lord is going to let it go on like this without some relief. But if we go into next fall without substantial storms, the whole West Coast could collapse, there could be starvation."

It is not that there haven't been warnings. As long

ago as 1973 the National Oceanic and Atmosphere Administration was predicting that "the probabilities of one or more drought years in the next three for soy beans is 7%; but for corn it is 26%, and for wheat it is 29%. The message is strong that we have been unusually fortunate in recent years to have experienced such high grain yields. It is imperative that we not be lulled into a dangerous and unjustified expectation that such fortunate circumstances will continue."

The Rise of Agribusiness

In the past twenty-five years agriculture in the United States and Canada has made enormous strides. Far removed from one man, one plow, one acre, modern North American agriculture is big business. And, as is often the case with big business, it necessarily involves other major facets of economic life in a network of symbiotic relationships.

First, it involves intelligence—what the weather is apt to be like (in the United States there are weekly crop bulletins, even daily ones for selected crops) and what prices different crops are apt to fetch in the marketplace.

Then, there's major investment in seed and stock, fertilizer and equipment. There's a continentwide system of storage facilities and a transportation system to move crops to their respective markets. There are processing facilities and, especially important for the farmers, there are price supports so that they have some certainty when they sow that they will reap at a profit.

If you move across America's wheat belt during harvest time, you seldom see agricultural workers in a field; what you do see is enormous machines. For modern farming is highly mechanized industry.

This has three consequences. The way land is worked in North America is exceptionally efficient in terms of using labor, that is, work per man-hour—

but (and it's a big but) it is not very efficient in terms of the land. No machine invented can shower the love and care on an acre bestowed by a rice-paddy farmer in Taiwan. Second, our agricultural production is tremendously inefficient in terms of our use of fossil fuels, already in short supply—oil, gasoline, and diesel fuel.

Finally, as farm equipment has become more efficient it has also become far more specialized. These modern machines have little ability to adapt to the difficult tasks involved in different new crops—a problem hands do not have.

Now, should the weather turn truly "normal," the dislocations it creates would be enormous. All the elements in our large and complex agricultural system are based on what is now a false assumption: that weather, and therefore crop yields, will operate within fairly narrow limitations. We have now enjoyed two exceptionally good decades—but can our agricultural technology cope with weather that won't conform to what we have come to call "normal"?

And this becomes even more worrisome when, as the CIA puts it, "it seems clear that the world of the poor, at least, will experience continued food shortages and occasional famine over the coming decades . . . the developed countries can expect to remain well fed, though perhaps not to raise their consumption of grain-fed animal products as fast as they might want to." (See Appendix 2.)

"Less-developed countries" (LDCs), "Third World," "emerging nations," and other euphemisms all mean the poor, the have-nots of the world. Today, these nations contain seven out of every ten people on earth, and their majority increases every day—"the disparity between the rich and the poor is," says the CIA, "likely to get even wider." They have the numbers, we have the grain and the goods.

We no longer risk antagonizing them—we already have. In their view, both history and our current practices are riddled with injustices. Some of the principal problems, past and present, can be simply stated:

1. Build an empire. Keep its people in relative servitude by capitalizing on existing feudal institutions. For example, keep India's moneylenders in place—it keeps the natives quiet.

2. Exploit the resources of each part of the empire, but keep modern technology and manufacturing at home.

3. As the empire begins to break up, prolong its existence by discouraging investment in the resources that might unify the country—modern transportation and communications systems—and in areas designed to provide self-sufficiency, such as education and agriculture.

4. Once independence has been granted, make long-term planning virtually impossible by letting basic commodity prices (copper, zinc, tea, coffee, rubber —these countries' major source of foreign funds) fluctuate with world demand based on industrial economies.

5. Make sure that foreign aid is highly selective so that it rewards our "friends," irrespective of their repressive measures (after all it takes strong measures to be absolutely reliable), and punishes our foes.

6. When aid is given, don't aim to make agriculture self-sufficient. Instead, make sure it goes into high-profile, high-technology areas (forgetting that high technology generally uses few people—the one asset LDCs have in abundance), which will compare favorably with offers from the competition. This is sometimes referred to as the "my dam is bigger than your railroad" theory of giving.

Thus, for example, until 1969 America's AID program specifically prohibited giving aid that could lead to increased land productivity. These aid limitations resulted from the politically oriented fear of being seen to create overseas competitors for U.S. crops.

7. Assist developing countries in keeping grain

prices at artificially low price levels to keep the urban poor quiet. This has two predictable consequences: By destroying the prospect of a fair market price for agriculture, it destroys the peasant farmer's incentive; it encourages absentee landowners either to produce inefficiently or to withhold their land from cultivation entirely.

8. Having assisted in the destruction of any economic basis for agricultural improvement, lecture sternly about birth control, disregarding the fact that infant mortality is high and children are seen to be an asset—to help with the work now and provide support in old age.

While all of the above has been going on, the rich nations have been using irreplaceable natural resources, notably fossil fuels, at a rate that guarantees their relatively prompt disappearance from the face of the earth. This adds to the poor countries' frustrations; if they ever catch up, there will be nothing left.

The result throughout the Third World has been migration from the land to teeming cities. The poor come in despair to try to find work, education for their children, and medical care. It is in the cities that they can see for themselves the disproportionate wealth of a tiny privileged class, and of the tourists. In some cases their feeling of inequality is reinforced by exposure to the world's most powerful selling medium—television—in public places, where they can see a lifestyle they'll never have.

Implicit in this situation is the potential for urban violence, which has forced Third World politicians to concentrate funds on temporarily pacifying the cities —so much so that the vast majority of external economic aid has been spent in urban areas. Much of this has gone to subsidize food prices at artificially low levels. A pittance has been devoted to attacking the root of the problem: low agricultural production.

One of the CIA reports highlights the problem: "The political commitment to agriculture has thus far been lacking. In most LDCs the governing policy has been either to ignore or to soak the peasants in order to promote industry and keep the city dweller reasonably content." To reverse this policy, the report says, would require enormous sums of money and skilled workers.

Although we have sent our agricultural experts to underdeveloped countries for many years, much of our assistance has been inappropriate. American agricultural experts, while full of good intentions, have often found themselves helpless without the scientific and technological backup routinely available to them in the United States. We learned painfully and at great expense that what works in Kansas will not work in Karachi.

There were bright spots; certain things have worked. Beginning in Mexico, Norman E. Borlaug, funded by the Rockefeller Foundation, began to develop specific strains of grain that were ideally adapted to local land and weather conditions. This gave rise to enormous increases in productivity per acre—the Green Revolution had arrived.

But now we know that the world's weather is changing. These specialized strains of grain simply may not work under coming weather conditions. The Green Revolution may prove to be a slender reed indeed, not strong enough to keep millions from starving.

The effects of weather change on man's ability to produce enough to feed himself are beginning.

In Asia each arable hectare (about 2.5 acres) currently feeds 3.5 people. If the average temperature were to drop just 1.8°F., that same land could support only 2 people. The figures for China are even more dramatic—an average 1.8°F. drop in temperature would plummet the land's potential yield from 7 to just 4 people per arable hectare.

As one CIA report points out, available land is already scarce in Asia, and what there is is intensively farmed. In the world as a whole, more and more farm land is lost each year, through careless farming, which causes erosion, as well as through industrialization and urbanization. "Clearly the greatest potential for in-

creased food production over the longer run lies in
the LDCs," says the CIA report, "where yields are
far below those of the developed countries." The
social, political, and economic obstacles are formidable,
and adequate incentives and assistance for farmers
imply a "major shift in the rural-urban terms of trade
in most LDCs. If food prices paid to farmers go up,
the urban poor cannot afford the increase. Either they
get subsidized food or starve."

Many of us perhaps think of the Third World as
jungle, bush, and small villages. We'd be nearer the
truth if we conjured up a picture of a sprawling, fester-
ing slum. The Third World's major cities are big and
growing two to ten times faster than those in the
industrialized world. Some major ones and the currently estimated population in millions for each:

Bangkok	4.2
Bombay	6.2
Buenos Aires	9.0
Cairo	5.3
Calcutta	7.2
Dacca	2.0
Jakarta	4.8
Karachi	3.7
Kinshasha	2.1
Lima	3.5
Madras	3.4
Mexico City	11.0
New Delhi	4.1
Santiago	3.1
São Paulo	7.9
Seoul	5.8
Taipei	2.0
Manila	3.2

We, the "civilized" world, have created the un-
civilized food bind in which the less-developed countries find themselves enmeshed today. It was not all
our fault. America had fewer client states than most,

The United States is proud of its foreign-aid record. Yet, compared with many other governments, U.S. overseas aid is, relative to our domestic wealth, miserly. Great Britain, with all its economic problems, gives a greater percentage of its wealth away. The Netherlands and Sweden each give away almost three times as much as we do. The following illustrates this:

Foreign Economic Aid given by Selected Countries 1975 (Latest Year Available)

	Total Aid	Aid as Percentage of Gross National Product
United States	3,731.0	0.25
United Kingdom	777.8	0.34
Canada	878.5	0.58
Australia	507.9	0.61
Germany	1,526.4	0.36
France	1,979.8	0.73
Netherlands	583.8	0.73
Sweden	563.1	0.81

Put in personal terms, our aid looks like this: If you made eighteen thousand dollars a year and gave at the rate of the U.S. government, your total donation to all charities would be forty-five dollars. If you gave on U.S. government terms, you'd insist much of this was paid back to you in purchases (often of things you'd otherwise have to store — like grain).

and the Philippines hardly constitute an empire—although many plausibly argue that Latin America has been our hidden fiefdom for far too long. As Porfirio Diaz put it: "Poor Mexico: So far from God, so close to the United States." Britain was, on balance, a fairly just colonial power. So, in certain ways, was Holland.

And the less-developed countries themselves are not universally governed by wise and beneficent rulers. In fact, normally quite the reverse is true. Freedom House puts out an annual list of countries where, in their judgment, totalitarian practices prevail. It grows each year; today only 19.6 percent of the world's population is considered truly free.

But the hungry man in the overcrowded city does not care who was or is his theoretical master. He knows only that he must feed himself and his family now. In a world of uncertain climate, that's going to be an ever more difficult task.

7

The Overflowing Shopping Bag—
Energy, Consumption, and Waste

The Wisconsin group predicted that the climate could not support the world's population since technology offers no immediate solution . . . Moreover, they observed that agriculture would become even more energy dependent in a world of declining resources.

—CIA report

The average American, like a heroin addict, is a junkie. We are high on energy, a habit we can no longer afford. For a long time our habit was supportable. Energy was cheap. We took it for granted. We had easy access to it. It had, in short, become our right.

The climate and the social climate were also right. America was the ultimate society of abundance.

There were many pushers encouraging our habit. Manufacturers stridently told us that we could not do without electric can openers and refrigerators that spit out ice cubes automatically. We believed. Detroit urged us to "See the U.S.A. in your Chevrolet," and made sure that we knew that big equaled beautiful. We listened. The oil companies offered endless free glassware, steak knives, and the chance to match money and make a fortune, all in an effort to see that we guzzled gas. Then there were the utilities themselves, which promoted the kilowatt, nicknaming it "Reddy." Most important, "Reddy" was cheap.

Not any longer. The past winter put an end to that. With strains on the fuel supply increasing—and with uncertain, unpromising weather in the future—"Reddy" will never be cheap again.

In some ways, being an energy addict is different from being a heroin addict. With heroin, you start small but can end up first mainlining, then dead. With energy, in the mid-twentieth century, America and the world have been more creative, less straightforward.

Take the matter of packaging. It has been refined to such an art that often more energy goes into production of the package than into the product it contains. And we do not recycle nearly enough. Aluminum illustrates this point to perfection: *It takes twenty times as much energy to make one pound of new aluminum as to melt down aluminum scrap.*

Our cars are packages, too—mammoth boxes weighing thousands of unnecessary pounds. In a nation that prides itself on efficiency, they are monuments to inefficiency. There are over 109 million of them on the roads today. They average less than 14 miles to the gallon. But then, as Henry Ford put it when asked why Detroit did not push small cars, "Well, you can't make a profit on a small car as large as you can on a large car, simply because the price you sell a large car for is a lot more than the price you sell a small car for—so the profits have got to be smaller."

Then there's the matter of innovation. Since the 1950s, much of our industry—and therefore much of our energy—has gone toward producing things we never knew we needed: electrical pencil sharpeners, snowmobiles, ever-more-powerful (and power-using) hairdriers, snow blowers, electronic games, photocopiers, CB radio sets . . . The list is nearly endless.

And, if the list is big, so is the energy bill. Today, experts estimate that every American requires the daily energy equivalent of 8 gallons of gasoline. Further, they predict that if we don't get off the energy binge, this will grow to 3,696 gallows per year per American by 1985, two and a half times our rate of use in 1950, 50 percent again as much as in 1970.

Energy—The Silent Servant

It is easy to explain this amazing rate of consumption. We use energy because it is there, because we've grown accustomed to it, and because our memories are short.

Were they longer, we would remember that in our recent past the rules were different, more in line with reality; that, as Ralph Waldo Emerson put it, "Nature never gives anything away. Everything is sold at a price. It is only in the ideals of abstraction that choice comes without consequence."

If we wanted to heat a room, we put logs on the fire, logs that had been cut by hand, or we shoveled coal. If we wanted light, we lit a candle or filled a lamp and trimmed the wick. If we wanted something from the marketplace, we walked to get it or we saddled up. In short, all of our actions required physical effort, and that very effort itself limited consumption. We learned what we could do without by knowing what wasn't worthwhile.

Now, with limited resources, as we change from a social climate of abundance to one of scarcity, we must relearn those most basic lessons. Everything has a price. Life is a series of hard choices.

The Winter of Our Discontent

The harsh realities of the critical interaction of climate and energy were amply illustrated by the winter of 1977. During the storms, gas consumption went from 57 billion cubic feet daily to 100 billion. As temperature plunged, so did fuel availability. At one point natural gas was down to a three-day reserve.

In Pennsylvania, the situation was so desperate that garbage was burned to make gas, and an old power plant—so old that it was being turned into a museum—was reopened.

Frozen rivers and canals trapped millions of barrels of oil and liquified gas in transit to freezing customers. The winter's weather was so bad that our oil imports boomed to 50 percent of total consumption.

After forty-five days of subzero weather, Buffalo's head weatherman for twenty-six years quit, saying, "After all this, I just ask myself, what am I doing here? This winter was the last straw."

Now it is over, but the aftereffects remain. The Federal Energy Administration estimates that the average American's heating bill will be up by 35–45 percent as a result of the weather, and Charles L. Schultze, President Carter's chief economic adviser, believes the total cost to consumers could run as high as $5 billion.

Some Chilling Facts

The winter's weather highlighted America's dependence on scarce fossil fuels and foreign imports. Nineteen seventy-seven overall oil imports will exceed 40 percent of total oil consumption. The check America will be writing to OPEC countries will be for $40 billion—very nearly the equivalent of $200 for every man, woman, and child in the country. Put differently, our overseas oil payments would pay the salaries for over 5 million jobs or build over 600,000 new homes. This energy deficit is like any other deficit. It comes from living beyond our means. With only 6 percent of the world's population, the U.S. uses 35 percent of all energy consumed.

By any standard, our consumption is excessive. The average European family uses thirty-five hundred kilowatt hours of electricity a year in their home. We use seventy-five hundred. West Germany's consumption of gasoline per head, with the same pattern of car ownership, is only 20 percent of ours. In general, their housing standards equal ours, yet they only use half as much energy to heat them.

This profligate waste is near criminal, because the world is running out of fossil fuels. The U.S. Energy Research and Development Administration estimates that the world's proved total oil reserves will last for only 35 years, coal reserves for only 175 years of world consumption.

What Is Our Energy Situation?

"It was a gas," "We're hitting on all cylinders," "Let's have a blowout"—American slang is as full of energy-related terms as the "abundant" fuel that inspired them. Coal constitutes 90 percent of our short-term recoverable fuel reserves. Experts estimate that this may represent three times the energy potential of all oil in the Middle East.

Our proven oil reserves are no less than 40 billion barrels. Some geologists think continental-shelf drilling could more than double this amount.

Natural gas is sometimes a byproduct of oil drilling, and our potential natural gas reserves are thought by some to be vast, although the economic and environmental cost of developing them will be high.

Nuclear energy is surrounded by controversy, but there is one unassailable fact: We have an abundance of uranium right at home.

Why, then, did the U.S. almost run out of fuel in the winter of 1977? Why must the key to our energy future lie with countries whose names, till recently, were unfamiliar to us? What is our true energy situation, and what can we do about it?

A useful starting point is to look at where our energy supplies come from now and compare this with the past and with Western Europe's present.

Percent of U.S. Supply

	1920	1950	1976	Percent of 1974 Western Europe
Coal	78	38	19	34
Oil	14	39	47	19
Natural Gas	4	18	27	9
Hydroelectric	4	5	4	30
Nuclear	—	—	3	8

Today's situation is little changed from that prevailing at the end of 1974. In combination, oil and natural gas provide 77 percent of our energy requirements, while coal, which constitutes 90 percent of our reserves, now accounts for only 20 percent of our energy. Thus, in the last fifty years there has been a dramatic shift away from coal. Why this shift?

KING COAL

Extracting usable fuel, whether from coal seams or oil fields, takes money—lots of it up front, before you take any out. But coal has three additional disadvantages.

First, it is far more labor-intensive. It takes far more men to extract a ton of coal than it does to get its equivalent out of the earth in oil. And coal miners are expensive when compared with the work done by machines and by man in any other countries.

Second, strip mining (40 percent of U.S. production today) is vastly more efficient than underground mining:

	Strip Mining	Underground Mining
Capital Investment	$15 per ton	$20 per ton
Tons per Miner per Day	36 tons	12 tons
Recovery of Total Coal	90%	50–60%

But strip mining has enormous ecological problems. It mars the immediate area and affects the land for miles around. It can lead to serious flash floods. It rapes the land so abusively that it can take fifty years or more to recover, not to mention the fact that human poverty is attendant on land poverty.

Finally, there are two complicating factors: most of our developed coal facilities are in the East, but most mine reserves are in the West. If coal has over 1 percent sulfur content, it is unacceptable from an environmental standpoint. An estimated 65 percent of U.S. coal reserves have less than 1 percent sulfur, *but* only 40 percent of the coal we currently produce meets this standard, and much of this is earmarked for steel-making.

OIL

If coal has problems, petroleum seemed to have unlimited potential. Above all, it was cheap. And all of us, governments, oil companies, manufacturers, and consumers, went on a binge in the misplaced belief that we faced unending supply of cheap foreign oil. And cheap it was, and is. Even today, the true cost of a barrel of crude oil from the Middle East (the foundation for twenty gallons of gasoline plus twenty-two gallons worth of oil products) is less than fifty cents delivered.

Low price also meant a low return on investment with a relatively high risk (only 35 percent of U.S. wells ever come into production). So crude-oil exploration declined in the U.S. throughout the sixties, while it boomed elsewhere in the world. The major oil companies concentrated on where they knew they'd strike it rich (be that Venezuela, Nigeria or Iran). Different petroleum end products require different ships to carry them across the high seas. That's expensive. So instead, the oil companies made the decision to transport crude oil and refine it at home ports, close

Proven World Oil Reserves

	Billion Barrels
Middle East	340
Communist Countries	65
Western Europe	25
Western Hemisphere excluding U.S.	42
U.S.	40
Africa	60

to consumer markets. The age of the supertanker and of super ports, such as Rotterdam, had arrived.

Throughout history, when man has an abundance of any commodity he has tended to waste it because he views it cheaply. U.S. postwar society was predicated on inexpensive oil. Our electrical generating plants were built to its specifications. So, too, our cars, homes, and many other aspects of our daily lives were constructed with little regard for fuel economy. A series of other factors operated as well.

1. After the Santa Barbara oil blowout, we temporarily cut off continental-shelf exploration.

2. We found enormous reserves on Alaska's North Slope, then delayed construction of an eight-hundred-mile trans-Alaskan pipeline because of largely justified environmental concerns.

3. Most important, in 1954 the Supreme Court held that under the 1938 Natural Gas Act the Federal Power Commission had to regulate not only what interstate pipelines could charge for the gas they sold to local utilities but also how much the pipelines could pay when they bought gas from producers (the wellhead price). In the years following, this latter tool was used to regulate interstate prices at artificially low levels. As a result of this underpricing, the best fuel we have from an environmental standpoint is woefully over-used—and misused as well, in such areas as industrial power generation.

Finally, our development of alternative energy sources moved forward at a snail's pace because, for one thing, extracting gas or oil from coal (syngas and syncrude) is enormously expensive. And, for another, we assumed that nuclear energy would be the ultimate answer, so we didn't look elsewhere. Those in the business of atomic energy used advertising and public relations to assure Americans that nuclear energy was the wave of the future.

Now we know that plant safety, radioactive waste materials (it takes thousands of years for some waste to lose half its radioactivity), and the possibility of terrorists' getting their hands on radioactive materials pose major problems. (The government now requires nuclear power plants to enforce security measures that theoretically would be adequate to cope with a conspiracy of two or more persons inside the facility who were acting in concert with an outside group of terrorists armed with automatic rifles, recoilless cannons and plastic explosives.)

The Morning After

As with all binges, there was a morning after; the Western world suddenly woke in October, 1973, to face an incredible hangover. The Arab world was strangling us by withholding oil. From October to the following March, 1974, when the embargo was lifted, we really had the shakes. Just 9 percent of our supplies were disrupted. It doesn't sound like much, but it was enough to throw as many as half a million Americans out of work, enough to cut our gross national product by an amazing $10 to $20 billion.

We had learned some hard truths: That you're not truly free if you're not energy-independent. That the OPEC countries could blackmail us.

That there was irony in our turning away from coal because of the price of paying more men, only to find that now we had to pay the price of one man. The bittersweet joke "Yamani or your life" entered the English language.

There were other implications. When the OPEC cartel banded together so effectively, it was politely suggested to the oil companies that they consider "participation" at source. Translation: "Give me your company."

There were enormous investments both inside and outside the OPEC countries with their newfound wealth. They spent on an incredible array, everything from jet fighters to jet-set hotels—Great Britain's Dorchester, for instance. The potential implications of this flood tide of money are awesome. The Saudis, the Kuwaitis, and some other OPEC states have the potential of inflicting damage on the banking system and international exchange rates by moving their money. The Arab States' reserves in Britain's pound are so great that, experts believe, were they to be entirely withdrawn, there is a real risk they would destroy the country's money. In fact, with the pound down from $2.20 to $1.70, it is already happening.

Finally, and most disturbing, with the exception of the major aspiring industrial nations (such as Iran, Venezuela and Iraq), some OPEC countries were wealthy enough to comfortably withhold all their oil for long periods of time. They had no short-term need for the money. Thus, they could quite literally destroy the economies of the Western world.

One couldn't really blame the OPEC countries. They owned a valuable and finite resource; together they had infinite possibilities. In the past, the oil companies, separately and as ARAMCO, had wielded power equivalent to any single state. Now things were different. The Arabs knew they'd never be the same again.

If OPEC was quick to realize its new-found power, the U.S. was slow to seek solutions that would provide viable alternatives. At the time of the embargo, we imported 36.1 percent of our oil. By 1976 it was 41.9 percent, and early in 1977 it was hovering close to 44 percent—a total of $93 million every day. And yet, less than thirty years ago the U.S. was an oil exporter.

The figures for other countries are of equal magnitude. The U.K., which spends $1.7 million per hour on energy, must currently import an incredible $17 million worth of oil per day—half of their total energy requirement. Fortunately for Britain, North Sea gas has now begun to flow. Many experts think that its late arrival is doubly lucky for that country. It will make an enormous difference to Britain's external energy bill by the 1980s. What's more, if Britain had had the oil earlier, they probably would have wasted it, as we did.

In the thirty years from 1950 to 1980, the world's industrialized countries' population grew by 25 percent. At the same time, U.S. energy growth was 400 percent. Our energy requirement has been doubling every fifteen years, and currently, with generally mild winters till 1977, we require 3–5 percent more energy each year. Since the embargo, OPEC has raised prices four times. Today a barrel of landed crude oil—cost: fifty cents—sells to us for over thirteen dollars.

We must afford these prices. We have no choice. But we cannot afford to continue our energy binge. From 1960 to 1965 our energy growth rate was 3.6 percent; then it increased by a third, to 4.8 percent, from 1965 to 1970. Now it is holding at just over 4 percent. The United Nations says that America, along with the rest of the industrialized world, must cut this back to 2.5 percent if we are not to face permanent shortage. Quite simply, if we cannot conserve, we will have no economic growth, no true freedom.

The Four-Hundred-Servant Household

In the past twenty years, the average private user has doubled his use of energy. According to the Natural Gas Association, if you take our energy and divide it by our citizenry, the amount of power each of us commands is the equivalent of four hundred people working eight hours a day all year. How we employ that energy in the trying times ahead is a critical question. America's current energy use looks like this:

How We Use Our Energy

	Percent by Use			
	Oil	Gas	Coal	Total
Electrical generation	10	15	66	26
Transport	53	3	—	25
Industry	18	50	32	29
Residential and commercial (light and heating/cooling and equipment)	19	32	2	20

As the figures show, gas is critical to industry and home owners. It heats 40 million residences, approximately 3.5 million commercial buildings, and is used in approximately 200,000 factories—many of which can't manufacture without it.

We waste 40 percent of our money. Why? What can we do about it and at what cost? A look at the individual parts of our energy bill begins to answer these questions.

What Can We Do About Wasted Energy?

ELECTRICITY

Generating electricity is a monumentally inefficient way to use fuel. A power-generating plant takes raw fuel, generates heat, and converts this into electricity. Even in the most efficient modern plants, only 40 percent of the fuel is converted into electricity. The rest is dissipated owing to the lack of storage capacity, inefficient generating capability, and leakage in power transportation. Waste of this kind means that nearly 15 percent of our total fuel requirements are squandered. This amount alone, if saved, would very nearly wipe out our oil deficit. And in certain other parts of the world the waste is even greater. For example, experts estimate that up to 70 percent of the oil Britain uses in its power plants is wasted.

The Carter administration is now providing incentives for utilities to switch from natural gas, which is in critically short supply, to oil, which may up our overall import requirements. By 1985, many utilities will have to switch to coal. This will represent the major step backward, since coal is a relatively less efficient fuel than oil in electrical generation.

The utilities themselves could conserve significant fuel if they put in better storage facilities. However, given the increasingly stringent pollution laws in the U.S., these savings might well be offset by the compensating cost/loss due to pollution controls.

TRANSPORTATION

The U.S. Department of Transportation estimates that as of the end of 1976, there were approximately 109.5 million cars on the road and 29.5 million trucks and buses. Little wonder that transportation accounts for one quarter of our total energy requirements. Transportation also involves major energy waste. Experts estimate that, depending on the vehicle, from 50 percent to 80 percent of the fuel is wasted.

In the past ten years, our gas consumption has increased by 50 percent. The reasons can be simply stated:

1. There are far more vehicles on the road.
2. The vehicles themselves use more energy due to increased size, air conditioners, power steering, and power brakes.
3. Mandatory pollution controls on auto emission

have reduced efficiency by 6 billion gallons per year. This equals 7 million barrels of gas per day—twenty days' worth of imports at current levels.

Many contend that our 55-mph national speed limit should be taken off. They're quite right—it should go down to 50. At that level, compared with 60, we would conserve one gallon in seven.

There are other ways that we are wasteful. Diesel engines are 28–35 percent more efficient, a fact Western Europe car owners take advantage of. In the U.S., however, they constitute a minute percentage of passenger vehicle engines.

We pride ourselves on being the most industrialized of societies, yet our mass-transit system is a joke. Commuter trains in Western Europe and Japan routinely carry passengers at twice our speeds. Amtrak has made some headway, but far more attention is needed. One rail line can move three times as many people as a three lane highway at a far lower energy cost.

America has potentially one of the world's best waterway systems. And a less expensive way to move goods has yet to be found. Even so, we have not done nearly enough to encourage the growth of this means of transit.

INDUSTRY

Inexpensive oil encouraged inefficiency in industry. Far too many plants have multiple sources of power with no provision for reusing any of it. Far more emphasis must be placed on enclosed power systems which store and reuse waste heat. This is expensive but can be done.

Certain types of industry use disproportionate amounts of energy. It is probable we will have to enact an end-use tax on energy input for industries that provide goods at inefficient energy levels relative to alternatives.

RESIDENTIAL AND COMMERCIAL

Twenty percent of our energy falls into this category. Home use accounts for about two thirds. The primary villains here are inefficient heating and air conditioning. Americans spend a disproportionate amount of their electricity bill air conditioning the outdoors in the summer and warming the outdoors in the winter. Experts estimate that up to one third of our homes now leak energy at an appalling rate—first, via poor insulation. Adequate insulation (measured in "R" terms, as explained in Appendix A) is bound to pay for itself in the course of two to six years.

In addition to leaking energy, we keep our homes too hot in winter, too cold in summer. This costs us and the nation enormous sums: the average family's bill for air conditioning and heating ranges from $500 to $650. For every degree we increase our thermostat above 70, our heating bill goes up 3 percent. Yours and the nation's.

Americans do, over the short term, respond to requests to conserve fuel. The Gallup survey shows that after President Carter's request to do so, 61 percent of Americans responded. Whether we will discipline ourselves to conserve via insulation and lower room temperatures over long periods of time is a different matter. Federal Energy Administrator John F. O'Leary doubts it. As he puts it: "We're going to have to find some way . . . to go back into all the buildings and all the houses that have been built, to bring them up to a minimum standard of energy efficiency . . . I don't think we can do that through voluntarism."

Nonsolutions

Any rational discussion on how to solve our problem must include the discarding of certain myths:

The problem will go away." The problem will not

go away. It will intensify. Currently we consume 18 percent more than we produce. By 1985, if this is not changed, this will rise to 25 percent. That's like writing a check for one hundred dollars when you have only seventy-five dollars in the bank.

True, we can export and make up the difference and our ability to export is prodigious. But we still are drawing funds against a bank owned by others. Oil is a political weapon.

The problem is all the more severe since we probably face colder winters for some time to come.

"Solar energy or nuclear power will bail us out." We are at least twenty years from having a widespread solar-energy system. The potential is enormous. The amount of solar energy reaching just seventy square miles of earth in one day could, if captured, provide enough total energy to supply the total U.S. annual consumption. That's the good news. The bad news is that we have not as yet perfected economic ways of efficiently collecting it, storing it, or transporting it.

Nuclear energy has severe problems. First, even if we wanted to proceed rapidly, we are woefully behind in our generating-plant construction program. Second, while the odds are exceptionally long on an industrial disaster, the consequence of such an event could be catastrophic. Third, there is the critical matter of waste disposal. The way we dispose of nuclear waste today could, if ill advised, plague subsequent generations for thousands of years.

This is not to imply that we should rule out the nuclear alternative. We cannot afford to close out any option at this point. But safeguards must approximate those that we have built into our missile system.

"Tax credits and other incentives will do the whole job." Offering home owners and industry a hidden benefit in the form of a tax credit for insulation, for shifting energy source, or other forms of conservation is an indirect way of attacking the problem. Historically, compared with direct measures, these have never worked. It is probable that we will have to enact

legislation requiring adequate insulation and, in the case of industry and utilities, specifying fuel selection and consumption.

"The oil companies have conspired against us. They must be broken up." There is every evidence that the oil companies have cooperated with one another for many years. Joint leases are one example. Another example, closer to home, is the way the oil companies, operating under their own initiative, allocated free-world oil during the embargo to ensure that our society did not collapse.

There are currently over ten different proposals before Congress to rip the oil companies apart. Some specify that vertical integration must stop; that the same company should not be allowed to produce, refine, and market. Others specify that the oil companies should be allowed to deal in only one kind of fuel. Both disregard fundamental realities:

1. The system requires gigantic long-term investment on the part of large companies.

2. Production, refining, transportation, storage and marketing are all intricately interrelated.

"Our standard of living will remain unchanged." This simply cannot be. For too long we have lived in what Governor Jerry Brown of California calls the "cowboy society." It is typified by the attitude: "If it's there, exploit it." Now we must reduce our standard of living in small ways—lowered room temperatures, fewer gadgets, smaller cars—and in big ways—legislated home improvements, energy-use taxes, and staggered working hours.

"Ecology must always take precedence." Today, we are understandably far more concerned for our environment than ever before. Long-term major pollution is unacceptable not only for esthetic reasons, but for practical ones. Government and society must together strike a realistic balance between ecological concern and the ability to sustain our country's economic growth.

For instance, it is probable that we must strip-mine for coal. There is no reason why we cannot do so under tightly controlled regulations and add an ecology tax to each ton extracted (experts estimate this three to six dollars a ton) to repair the damage done.

Oil and gas exploration on the outer continental shelf is another example. Some geologists and other scientists estimate that these holdings may contain as much oil as the U.S.A. has used in the last one hundred years. The Santa Barbara oil spill illustrated the dangers of offshore drilling, but, given adequate safeguards, we can protect against a recurrence. Nineteen thousand offshore U.S. wells have been drilled. They have yielded 4 billion gallons of oil and 33 trillion cubic feet of gas. There have been only four blowouts that caused serious pollution.

"The energy producers should produce uneconomically." The United States is a capitalist system. Our economy has thrived on the profit motive. Energy producers, along with all other American businesses, exist to make money. As President Carter put it in his news conference of February 23, 1977: "If I were running an oil company, I would reserve the right to release or to reserve some supplies of natural gas." Beginning *now*, the price of natural gas and crude oil should be deregulated. Those prices would then find their own fair market price in our economy.

This would have two powerful, positive effects: It would stimulate domestic oil and gas exploration and production (in fact, this could be a required part of the deregulation), and the more expensive energy would bring home to every American the price of our waste. Consumption would automatically go down, as it has elsewhere in the world when the citizen has to pay the true cost-related price.

Toward a Realistic Future

Between now and the year 1990, it is probable that most of our energy supplies must be met by exploiting known existing resources and technologies: Alaskan oil, the outer continental shelf, more domestic oil and gas exploration, and increased mining of the vast coal resources we possess. At the same time, we should address ourselves to further developing the science of economically extracting oil and gas from coal and of oil from the enormous reserves of oil shale we possess in Colorado, Utah, and Wyoming. Geologists estimate these to be 1.8 trillion barrels of oil. They believe one third of this can be recovered. That's far more than Saudi Arabia's total reserves.

Geothermal energy, based on capturing the natural heat of the earth generated by its subsurface liquids, is another interesting potential energy source. One such plant is already in operation in Northern California.

Nuclear-power experiments, with adequate regard for safety, will probably play some part in any future energy mix.

For the long-term future, we must explore natural, nonpolluting energy alternatives. Solar energy, wind, and harnessing tidal and ocean currents are all exciting prospects.

America's energy addiction has caused us severe problems, but we must never forget that we are an enormously energetic and innovative people. Given discipline, government support, and public awareness, we will doubtless find a way.

8

Between the Rock and the Hard Place—The Politics of Famine

Climate is now a critical factor. The politics of food will become the central issue of every government.
—CIA report

"Triage" is a little-known term. You've probably never heard of it, unless you've been wounded on a battlefield or have been a medic under fire. Triage is brutally simple. When the wounded are brought to a field medical hospital, they're split into three groups (hence "triage"). One group will live whether or not they're treated quickly; they are temporarily ignored. One group will die irrespective of treatment, in the judgment of the person making the choice; they are left to perish. One group may live if treated at once; it is this group that receives immediate life-giving support.

Within the foreseeable future, according to the CIA, the U.S. will be in the position of making this judgment for a large part of the world's population. The cooling trend and the consequent climatic changes pose a threat to every man, woman, and child in the world. The CIA is not alone in its conclusion:

The direction of the change is such that if it persists, as it well may, we must expect almost certain crop failures within the decade. This, coinciding with a period of almost non-existent grain reserves, can be ignored only at the risk of great suffering and mass starvation (*"Climate Change and World Food Production." The International Federation of Institutes for Advanced Study, The Nobel House, Stockholm, Sweden, October, 1974.*)

If the world is going to have fluctuations in its overall food production on the order of 1 to 5 percent, and if that occurs at a time when we have poor food distribution systems, or inadequate reserves locally, then you run the risk of not being able to support on the order of 1 to 5 percent of the world's population. To give you some perspective as to whether that is a "catastrophe" or not, realize that the world population now is 4 billion people. A 1 percent change is 10 million people. You could, in 1 or 2 sustained, marginally bad food years, kill as many people as died in World War II. (*Testimony by Dr. Stephen H. Schneider, in front of Subcommittee on the Environment and the Atmosphere of the Committee on Science and Technology, U.S. House of Representatives, May, 1976.*)

America—Land of Plenty

Previous chapters have outlined how the climate is changing and the effects of that on the world. As the world returns to "normal" weather, the U.S. will continue to be singularly blessed. There are almost two hundred nations in the world. Of those, just seven are net food exporters. That means they produce more than they consume. Of all food exported, America provides more than 90 percent, as these CIA figures show.

The World's Imports/Exports of Grain

Net Exports (+) and Imports (−) in million metric tons[1]					
	1948-52	1960	1966	1972-73[2]	1973-74[2]
North America	+23	+39	+59	+89	+92
Latin America	+ 1	0	+ 5	− 3	− 2
Western Europe	−22	−25	−27	−18	−20
East Europe and USSR	0	0	− 4	−26	−12
Africa	0	− 2	− 7	− 1	− 5
Asia	− 6	−17	−34	−38	−49
Australia and New Zealand	+ 3	+ 6	+ 8	+ 7	+ 9

[1]Totals will not balance because of stock changes and rounding.
[2]Different series, but indicative of trend. 1973–74 is preliminary.

The CIA estimates that even under the most optimistic conditions, in the years to come there simply won't be enough food to go around. The situation is not new. It has long existed. The United Nations Food and Agricultural Organization estimates that of the world's current 4 billion population, 500 million people (one out of eight) are chronically undernourished—near starvation. But the growth of population, when combined with limits of food production, will cause a critical reaction—what Dr. Paul Ehrlich properly called a "population bomb."

By the year 1985, the world's population will have grown by 25 percent, to 5 billion. Optimists believe it will be 6.4 billion in the year 2000. Pessimists believe it may be closer to 8 billion. Either number is unsupportable, given changing climatic conditions and limited food supply.

If OPEC owns the oil, we own the food and food

will be more valuable than gold in the colder years to come. The Third World is already resentful. Our population growth is roughly one twelfth that of India. Yet, each new American consumes five times as much grain and sixty times as much energy as each new Indian. We tell them to cut down their birthrate. They reply that our profligate consumption equals their poverty-stricken one with twelve times fewer people. Both of us are right.

The realities of America's food weapon are already apparent. In 1976, our overseas sale of food stuffs was $22 billion, far larger than our overseas sale of weapons and support technology at $4 billion. And our government already acknowledges that food is a weapon. Numerous government officials have said that our grain surplus is a critical factor in the SALT negotiations.

The CIA believes that if the cooling trend is great enough to reduce production in areas such as Canada, the U.S.S.R., and north China and increase the frequency of drought in countries that depend on regular monsoons, U.S. surpluses would still cover the needs of most of the world, except in particularly bad years, such as 1972, when a number of major agricultural areas experienced bad weather at the same time. If food stocks were low when this happened, there would not be enough grain to supply both the less developed countries and the affluent countries. In the words of the CIA: ". . . if the weather is 'normal,' it is essentially the poorer LDC's that will become ever more dependent on U.S. food exports."

The CIA reports that the less-developed countries will suffer most. How much they'll suffer is a matter of conjecture. The United Nations Food and Agricultural Organization predicts that world food yields can increase by 2–3 percent each year, given "normal" weather. As already stated, it is virtually certain that we are going to have normal weather—that is, colder, more aberrant weather, as opposed to the abnormal weather we have had for the past fifty years.

Normal weather or not, no reasonable forecast of

food production can begin to equal population growth. Yet, U.S. food production can increase markedly. The U.S. Department of Agriculture predicts that the U.S. is capable of a 50 percent increase in feed grains by 1985 and a 33 percent increase in wheat. Conservatively, this would increase the value of U.S. exports to between $175 and $200 billion per year. Ironically, most of those who need our exported food will not be able to pay.

As their own crops fail, the underdeveloped countries will become ever poorer. Their crying need to be fed now will force them to divert investments from areas that could make them self-sufficient later to buying food overseas now—if they can get it.

The underdeveloped countries will have another problem in feeding themselves. Oil prices will predictably be much higher by 1985—they have, after all, risen four times in less than five years. Fertilizer and modern pesticides are oil-based, so their price will skyrocket as well. The poor countries of the world can rightly blame the industrialized countries for these price increases. We have driven up oil prices by our seemingly endless demand. Given the choice between developing self-sufficiency in agriculture later and feeding their population now, governments have no choice at all.

There are other ominous factors. Deforestation and land development programs in Southeast Asia and South America are not working. At one minute to midnight, man is learning that ostensibly sturdy rain forests are in fact incredibly fragile ecological areas. Cut down the trees and you destroy the land as well as affecting the weather adversely. For example, the Amazon River rain forests are a bank that replenishes itself. Moisture is deposited in the rain forest, then, under recurring weather changes, is released in the form of life-giving rain over wide areas—so wide that destruction of the Amazon rain forest complex would have significant effects on Western Europe's weather.

And yet the rain forests of the world are being hacked down at an alarming rate. One estimate is that they are being cleared at the rate of fourteen acres every minute.

Deserts constitute a huge part of the world's landmass, 57 million square miles—approximately 40 percent of all land—and they are growing at an alarming rate. The U.N.'s Food and Agricultural Organization (FAO) estimates that they have increased by 3.5 million square miles in the past fifty years—that is an area equal to the size of all Europe. This desert creep, or desertification, is as ugly as its consequences. It adversely affects food, fresh water, and fuel as it advances.

The growth of deserts is a worldwide phenomenon. The Sahara has grown by 400,000 square miles, an area approximately equal to Texas and California. Its southern border has grown by 60 square miles in less than twenty years. Chile faces similar problems; so do Pakistan, Russia, and China.

One solution that must be urgently pursued is conservation. We in the U.S. must build up our grain reserves. At the same time, grain importers—mainly the LDCs—must attempt to check population growth to keep demand down while stockpiling reserves.

This pressing need will be extremely difficult to accomplish. On the day that Richard Nixon was inaugurated as the thirty-eighth President of the U.S., our grain reserves (our world food bank as it were) stood at over eighty-nine days of world food supply. On the day that he left office in disgrace, the U.S. had a grain supply equal to just thirty-one days of the world's then-current requirement. Ultimately, historians may judge this to be the worst obscenity of the Nixon years.

How Will the Empty Plates Be Filled?

The Sahel and India already illustrate the problems we will shortly face. The CIA analysts put it clearly:

A substantially cooler climate could add new and powerful countries to the list of major importers and reduce Canada's exportable surplus. Among the most immediate effects would be rapid increases in the price of food in almost all countries, which would create internal dislocations and discontent. The poor, within countries and as national entities, would be hardest hit. What is happening now to the poor in India and in drought-stricken Africa is probably a pale sample of what the food-deficit areas might then experience.

In many LDC's, the death rate from malnutrition and related diseases would rise and population growth slow down or cease. Elsewhere, there might be waves of migration of the hungry towards areas thought to have enough food. The outlook then would be for more political and economic instability in most poor countries as well as for growing lack of confidence in leaders unable to solve so basic a problem as providing food.

For the richer countries, the impact would be mitigated, at least, by their very wealth. While standards of living in countries needing to import large quantities of food would probably decline, there would be little danger of starvation. Nevertheless, there would be varying degrees of economic dislocation and political dissatisfaction whose results would be very difficult to forecast.

Under coming conditions, clearly the rich countries of the world (in other words, the tiny minority) will have to make do with less food. With the exception of the U.S., they will run a food balance-of-payments deficit that will put a big dent in their economies and cause some degree of social disruption.

For the poor countries of the world, the implications are more ominous. Many will starve. This future will place an almost unbearable moral and political burden on the U.S. and its people. We must face up to the following "unthinkable questions":

1. How much food will we keep for ourselves? In other words . . .

2. How much are we willing to reduce our standard of living to feed others? For example, would we all be willing to stop using our air conditioners if the energy so released would feed another 400,000 people? Or would we rather stay cool and let them starve?

3. Are we going to give away our food or sell it?

4. Who gets how much, and why?

As the CIA puts it, "In times of shortage when food prices rise, it will be hard to decide how much should be reserved for domestic consumption, how much should be sold at high prices, and how much should be given in aid to the needy." The power to decide who shall live and who shall die is the ultimate weapon.

Climate does not respect national boundaries; production of food does. America is accurately termed the bread basket of the world. In mankind's coming quest for food, we hold all the cards. The Agency report states that the "U.S.'s near-monopoly position as food exporter would have an enormous, though not easily definable, impact on international relations. It could give the U.S. a measure of power it had never had before—possibly an economic and political dominance greater than that of the immediate post–World War II years."

In deciding who's going to get triage treatment, the U.S. will face some appallingly difficult ethical questions:

1. Should we reward countries who institute strict birth-control measures?

2. Can we dictate to people who have long regarded children as their only form of security in their old age and/or have religious beliefs which forbid contraception?

3. Should we punish nations such as India that went nuclear at the expense of agriculture?

4. Can we really starve individual people when it is their governments that have followed misguided policies?

The CIA puts it well: "In bad years, when the U.S. could not meet the demand for food of most would-be importers, Washington would acquire virtual life and death power over the fate of multitudes of the needy. Without indulging in blackmail in any sense, the U.S. would gain extraordinary political and economic influence. For not only the poor LDC's but also the major powers would be at least partially dependent on food imports from the U.S." (See Appendix 2.)

And, the Agency adds, where climate changes caused grave shortages of food in spite of U.S. exports, the potential risks to the U.S. would rise. The militarily powerful but hungry nations would make increasingly desperate attempts to get more grain in any wars they could. This is no laughing matter. China currently has an estimated 3,250,000 men under arms; Russia has 3,800,000 in its armed forces.

The threat of invasion is not the only physical threat we face. Massive migration, backed by force, could become a reality. There have been such migrations before; America welcomed a million Irishmen when climate caused the potato famine. Would we welcome a million now? Experts estimate that the failure of one Indian monsoon would cause 100 million to go hungry. Who would take them in?

These massive numbers highlight sobering facts in comparison with the previous Little Ice Age. Then the earth's population was small. Man could retreat singly and in small tribes as cold weather moved south. Sheer numbers militate against this today. The world would witness a stampede of unprecedented proportions. Until recently man has been an agrarian creature. Today, the poor gravitate toward the world's cities. The fastest-growing cities on earth are in poor countries —places like Bangkok, Buenos Aires, and Delhi. These urban poor have neither the physique nor the skills to survive as nomads in the wilderness.

Desperate Measures

"Nuclear blackmail is not inconceivable" says the CIA.

Seven countries are currently members of the nuclear club: France, the United Kingdom, India, China, the U.S.S.R., Israel, and the U.S. There is no question but that three of these, India, China and Russia, would be drastically affected by a new Little Ice Age. In a situation of famine, is it really so far-fetched to believe that governments would fail to reach for the nuclear trigger when death by starvation is the alternative?

Jimmy the Greek would probably tell you that the odds of getting "action" increase as the number of players multiplies. In the June 30, 1976, Second Annual Report of the Joint Committee on Atomic Energy, it was estimated that an added twenty countries could deliver the bomb within six years: Argentina, Austria, Belgium, Brazil, Canada, Czechoslovakia, Denmark, East Germany, Italy, Japan, Korea (North), Netherlands, Norway, Poland, South Africa, Spain, Sweden, Switzerland, and Taiwan.

Even if we assume that today's leaders are responsible and would not do so, a key question remains: Would today's relatively responsible leaders be in power?

Throughout history, food shortages resulting from bad weather and consequent harvests have caused radical social change. Louis XVI was deposed during the French Revolution as a result of food shortages caused by two years of exceptionally bad weather in France. More recently, Haile Selassie's long Ethiopian reign came to an end, mainly because of unrest caused by the drought in his country. Poland, a rigidly controlled Communist country, was forced to rescind food price rises when people rioted in the streets. Sadat faced a similar situation when food related riots erupted in Cairo. By giving in to the rioters, he has mortgaged part of his country's ability to become self-sufficient in industry and agriculture later. The other Arab states have had to bail him out with over $1 billion in assistance.

Angry, hungry people are not logical. They are looking for scapegoats and so can easily be exploited by demagogues who promise simplistic solutions in a complex world. As the CIA puts it: ". . . rural masses may become less docile in the future and if famine also threatens the cities and reduces the living standards of the middle classes, it could lead to social and political upheavals which cripple governmental authority. The beleaguered governments could become more difficult to deal with on international issues either because of a collapse in ability to meet commitments or through a greatly heightened nationalism and aggressiveness."

The CIA believes that there could well be ill-conceived efforts to undertake drastic cures. These might well be worse than the disease—changing the climate by trying to melt the Arctic icecap, for example.

Tampering with the weather could be the ultimate folly. Nature follows a basic law in physics: Every action has an equal counteraction. But desperate men are driven to desperate measures. For example, some anthropologists believe the Aztecs may have sacrificed their children to gain favor with weather gods. In the words of the CIA, "The potential for international conflict due to controlled climate modification can be a reality in the 1970's." One of the agency's tasks will be to trace, and anticipate, another country modifying the climate to its own advantage—and the detriment of the United States. Scientists the world over know many things they can do to change the climate. They can *estimate* what the short-term effects will be—but they are in hopeless disarray as to what would occur long-term.

The bleak choices the U.S. might face over who in the world will and who will not receive food in the future are put in large focus now at home. The western states are gripped in a drought of unprecedented proportions. At a meeting in Denver, western governor after western governor pleaded his state's case. They did so in the face of a bleak reality. Come spring, the Colorado and other western rivers will fall far short of supplying adequate water for all. This poses another set of unpalatable choices:

1. How much should a state like California get as opposed to a state like Arizona?
2. How should water be allocated within a state? For example, is it better to allocate water to labor-intensive crops such as lettuce and thereby stave off hardship and unemployment, or to agribusiness and thereby make the most efficient use of available water?
3. When does our good-neighbor policy end? Northern Mexico's agriculture is critically dependent on U.S. water. At a time when we are trying to redress years of neglect toward our neighbor to the south, just how far can we turn down our tap?

Preparation Is the Key

For some time, America has given lip service to the need for major coordinated international efforts to cope with climatic change. As far back as 1974, Henry Kissinger stated that "the poorest nations, already beset by manmade diseases, have been threatened by a natural one: the possibility of climatic changes in the monsoon belt and perhaps throughout the world."

Kissinger then committed the U.S. to a major investment in climatic research. Today, there are seventy-two weather-modification projects underway. Yet the total amount spent per annum is less than $40 million —less than we've spent on the development of the B-1 bomber in one month.

The CIA believes we must assume that the worsening of our weather will reach the point where even the best efforts of the U.S. would not be enough to meet the minimum needs of starving countries. But we have done nothing to act on such an assumption. Like children, we cover our eyes and hope the menace will go away. It will not.

The weather weapon exists. It's already in use. Poison gas used in World War I was shelved for a practical reason: It blew back on the user. Since Nagasaki, we've refrained from the "nuclear alternative": It too can boomerang. As we enter the final quarter of the twentieth century, the United States, with less than 6 percent of the world's population, must recognize that the short-term power our food weapon will bring can lead to long-term disaster. With common sense, we will.

9

Take Your Chances—Some Scenarios for Survival

In the poor and powerless areas, population would have to drop to levels that could be supported. Food subsidies and external aid, however generous the donors might be, would be inadequate. Unless or until the climate improved and agricultural techniques changed sufficiently, population levels now projected for the LDCs could not be reached. The population "problem" would have solved itself in the most unpleasant fashion.

—CIA report

Given the coming uncertain weather and the enormous food problems we face, what can we do? The Food Policy Research Institute predicts that, using the last seven years as a base, the less developed countries will run a food deficit of approximately 200 million tons by 1985. Where are we going to find the food to meet this coming demand?

Broadly speaking, the scenarios for survival break down into three distinct categories: What should be done internationally? What should be done domestically by developed and underdeveloped countries? What should each one of us do to make our own contribution to the fight against famine?

123

Major international cooperation has always been difficult to bring off, but there are reasons to hope that governments will sublimate their short-term interest to the pressing need for solutions to avoid worldwide disaster. Ample precedent exists for unlikely alliances among sovereign states when confronted by major outside menace—Britain and the United States and Russia, for example, during World War II. Now we have very possibly the greatest outside menace the world has ever known. Weather is no respecter of national frontiers; all may suffer in the years to come. Given these circumstances, surely all but the most xenophobic of national leaders will see the need for urgent cooperation.

First steps have already been taken in that direction. The United Nations World Meteorological Organization (WMO), with 143 member nations, is currently coordinating the largest weather-research program ever undertaken. Called World Weather Watch, it involves over one hundred countries that pool their meteorological data to give us all a clearer picture of the weather and of what creates different climatic conditions.

Both the WMO and individual nations' research programs into climate require massive coordinated funding to: 1. give us a far more extensive early-warning system for weather changes—especially necessary in the Southern Hemisphere; 2. begin to pinpoint more accurately man's effect on climate and how this can be kept under control; 3. develop the computer technology to process alternate programs in minutes instead of days. This is particularly vital. If one single factor had to be isolated as the key to our space program's success, it would be the highly sophisticated use of computers in designing, testing, and controlling all aspects of space flight. Today, climatology requires the same sort of intensive commitment in men, money, and machines.

Another important area for international cooperation is oceanography. Our seas are 75 percent of our earth; many suspect they could provide near-proportionate amounts of food. Although the oceans are an obvious resource, they present major problems. Pollution and overfishing have seriously cut into the world's catch. For much of the past ten years it has actually declined, so that today the world's oceans only yield between 65 and 75 million tons each year. The United States' recent adoption of the two-hundred-mile limit, along with its enforcement elsewhere, will make stricter conservation measures possible. However, the question remains: Can anyone afford to have national fishing beds when worldwide famine threatens?

The problems this poses are highlighted by the increasingly large number of governments that are insisting that all research within their respective two-hundred-mile limits be conducted by themselves.

The last international landmass, Antarctica, may provide a partial solution. As its surface waters are drawn northward by the earth's rotation, deeper, warmer waters come to the surface. This upswelling water, incredibly rich in nutrients, in turn feeds what are known as krill, small protein-rich crustaceans. They are in abundant supply; some scientists estimate that we could harvest from 100 to 150 million tons of krill, double our current fishing catch.

However, there are snags. The least of these is the taste of krill, which is abysmal; it could probably be disguised. More serious is the fact that whales and other sea life depend on krill for food. Major catches could disrupt the natural interdependent life cycle of the ocean. And, as has been shown, the consequences of tampering with the world's ecosystems can be severe.

Major international funds must be directed toward oceanographic research which will lead to intelligent, planned farming and harvesting of the world's vast ocean reserve. Both the United Nations Food and Agricultural Organization and UNESCO are already engaged in this work. Again, an international framework already exists.

International trade patterns and practices must be radically altered to assist the poor world in their drive to become agriculturally self-sufficient. Most of the cash crops the underdeveloped world relies on—coffee, tea, tobacco, rubber, hemp, and jute—are, from a protein standpoint, worthless. But exported, they earn cash to import food. We will have to guarantee these commodities at fixed prices in order to allow the LDCs to take some of their land out of cash-crop production and put it into grains.

In the industrial area, we currently keep the LDCs in a permanent state of dependence on raw-material exports by charging tariffs at a higher ratio as they export finished or semifinished goods. Thus there is a positive disincentive for moving from unproductive crops to industrial goods as a source of foreign income. This situation must be changed—possibly via a two-tier tariff system, with lowered rates for the poor countries —to encourage intelligent investment in light industry, transportation, communications and, in the longer run, agricultural self-sufficiency.

Fortunately, the world now has a food bank; it is the International Fund for Agricultural Development (IFAD). IFAD starts with a total of $1 billion pledged by thirty-four different countries. Interestingly enough, the OPEC countries have given over $400 million of this total, and thirty developing countries including OPEC, are members.

The Fund will provide monies for agricultural projects aimed at increasing food production in the developing world at bargain rates and, importantly, two thirds of the votes as to how the monies are invested lie with the developing world. Much like the World Bank, IFAD should be internationally administered and should hold backup food reserves, which it would allocate, like any banker, according to the responsibility of the lender. It should be empowered to ask such tough questions as:

1. What is being done to remove land from absentee ownership and place it under farmer owner control?
2. What international standards of grain storage can be guaranteed? Many experts believe that the

amount of grain which rots or is infested because of inadequate storage could more than make up any deficit for the foreseeable future.

3. What guarantees and supervision can be built in to insure that grain supplied reaches the truly needy without either yielding middlemen an excessive profit or disrupting the stability of domestic agricultural prices?

4. What is being done to encourage domestic agricultural production?

Recognizing that overgrazing on the Sahel can affect the United Kingdom's weather, and that destroying a Brazilian rain forest can affect the weather in Birmingham, Alabama, governments must cooperate in identifying, then moving to protect, selected fragile areas around the world. It is probable that only by supplying aid on an unparalleled scale can the rich nations of the world assist the poor in reordering land development priorities and relocating the people this would involve.

Is this suggestion hopelessly idealistic? Possibly. But nations have sacrificed more in the past in the face of a common enemy. Now we face potentially the greatest enemy of all: nature changing in unknown ways. And climatologists of all persuasions are very nearly unanimous on one point: Our depredations on land today may have unimaginably bad effects on our weather tomorrow. Can nations afford not to combine now? The United States itself has a variety of important options. In terms of foreign policy and foreign aid we must radically revise our thinking.

Continuing to give our food away overseas is not a solution. This simply cements a pauper relationship between the LDCs and the U.S. It does not attack the basic need for improving agricultural yield and agrarian land reform. America must address itself to a massive program of agricultural and economic assistance to increase yields around the world. We have the money to succeed. In fact, after World War II our economic aid as a percentage of our gross national product was almost nine times what it is today.

As part and parcel of that aid we should insure that local farmers get a fair market price (not a depressed price due to overseas subsidies). Only with proper economic incentives will supply and demand begin to move into equilibrium. Only by making it profitable to work the land will we stem the rising tide of migration to the poor world's cities.

Domestically, there is much that the U.S. can and should be doing: massive work on the creation of new grain strains, resistant to each of many likely new weather conditions; intensive development of hydroponics—fish-farming. Cold-blooded fish convert meal to protein far more efficiently than animals do.

We must also learn to conserve our water, along with sunshine, the foundation for growth of all living things. Currently, in certain parts of America, such as the Southwest, we are taking five times as much water out of the earth as nature, our climate, returns.

We also must learn to use water more prudently in other ways. Poorly thought out irrigation schemes cause an enormous amount of soil erosion each year. Some authorities put the number of acre equivalents lost through erosion at close to 1 million each year.

We need to give major attention to microculture— the intensive development of food in small spaces, including crops grown in combination with solar heating.

The United States government must level with its citizens, and explain that all man's reserves are finite. Given the earth's natural limitations, our current phenomenal rate of waste, inherent in our current consumption of both fossil fuels and food, must stop.

To partially correct fuel waste, major investment is needed in structuring an adequate mass-transit system. In addition, it is probable that taxes should be based on the horsepower and corresponding energy efficiency of automobiles. Energy prices must be raised to reflect America's energy import bill and the scarcity value of fossil fuels. The increased revenue thus raised should not go to the energy companies as windfall profits, but rather should be used in a variety of imaginative ways to foster research into alternative fuel and food sources and more efficient use of those we already possess.

By the same token, private dwellings should be taxed according to the efficiency of their energy use.

People who car-pool would be rewarded, as would those whose housing designs and insulation were energy efficient. The latter may require federal regulations.

Finally, the government should consider imposing a food tax based on the nutritional value of the food we eat. If you want to eat a pound of choice steak, that's fine—but the price should reflect the fact that you are consuming the equivalent of thirteen Asians' daily diet—sixteen pounds of grain.

In the seconds it is taking you to read this paragraph, people are starving all over the world. In the unsettled climatic conditions to come, it is probable that this bitter reality won't vanish; instead, the number starving each and every minute will likely increase. In the circumstances, America must provide positive incentives for our agriculture to increase grain for human consumption and decrease animal grains and the waste they involve.

As Frances Moore Lappé puts it in her thought-provoking book *Diet for a Small Planet*: "the amount of humanly edible protein fed to American livestock and not returned for human consumption approaches the whole world's protein deficit!"

Above all, we must recognize that if urgent action is not taken *now,* millions will starve in the near future. To disregard this coming reality is, quite simply, morally unacceptable. Americans, our government, and all the world's governments must recognize that changeable weather and the food crisis that will accompany it is not just a world problem. It is *the* major problem in the world today.

10

Energy-Saving Tips

Your Car

There are over 100 million cars on the roads in America today. On average, each year they travel about 10,000 miles and consume 700 gallons of gasoline. In total, they consume 14 percent of all our energy requirements. That's a lot. Here's how you can help to cut that total:

BUYING YOUR CAR

1. "EPA/FEA Gas Mileage Guide for New Car Buyers." Your first move when buying a new car should be to the post office. Send for the government's free guide, which gives you all the fuel-economy facts. The address is: Fuel Economy, Pueblo, Colorado 81009.

In selecting your car, and particularly a second car, try to figure out what is the most economic vehicle you can get by with. You don't need two station wagons.

2. Radial tires. These are markedly more efficient than non-radials. They're also safer.

3. Stick-shift versus automatic. A manual shift is markedly more energy-efficient.

4. Light-colored cars. If most of your driving will be done in a warm climate, light-colored cars are a good idea. They absorb far less heat, and you save on air conditioning.

5. Power accessories. Car air-conditioning, power brakes, power steering, power windows, and so on eat up energy. By cutting these down, you save money on the purchase price—and on every day you drive.

SERVICING YOUR CAR

1. Regular tune-ups. Make sure your car goes in for a checkup regularly. Experts estimate this can save 10–25 percent on your fuel bill.

2. Clean air filter. This can add up to two miles to the gallon.

3. Follow the manufacturer's recommendations on gasoline octane and oil grade. Using too high an octane wastes money and won't give you either better mileage or more power. Using an oil with a higher viscosity than recommended tends to increase friction and decrease fuel economy.

4. Tire pressure. Underinflated tires create more surface friction and thus require more gasoline.

5. Wheel alignment. Have your wheels checked regularly for proper alignment. Front-end wobble cuts way down on mileage.

6. Pump savings. Wait till your car is three quarters empty before refilling—gas adds weight, and weight reduces economy. Also, make sure that the gas attendant doesn't overfill your tank. Spillage is a complete waste.

PLANNING YOUR USE OF THE CAR

1. Car pools. One third of all gasoline is used in commuting. Yet too many cars contain only one person. Pooling saves everyone's gas and your money. It also cuts down on traffic jams.

2. Consolidate trips. Plan ahead so that when you are driving to do chores, you accomplish a lot all at once, as opposed to making a separate trip for each one. This saves you time and saves the nation valuable energy.

3. Preplan longer trips. If you are going to take a

long journey, it is worthwhile to get out a map and work out different routes. This could save substantial time and money.

4. Consider alternatives. Trains, buses, and planes are all more fuel-efficient. You'll arrive relaxed, while you've saved on the nation's energy bill.

UNDERWAY

1. Start only when you are ready. Wait until the car is fully loaded before starting the engine.

2. Accelerate to top gear smoothly. Abrupt changes are wasteful.

3. Drive at a steady pace—forgo lane jumping or "gunning it" in favor of steady, consistent driving at steady speeds. This is the way professional economy drivers drive—so should you.

4. Minimize braking. Abrupt braking and accelerating waste gas. Instead, look far enough ahead to anticipate, then use your accelerator to adjust your speed according to conditions.

5. Cut idling. If you are stalled in traffic for over a minute or waiting for someone, it's better to turn your engine off than keep it running over sixty seconds. Thereafter, it's far more energy-efficient to restart the car.

6. Use energy sparingly. You will get 21 percent better mileage at 55 mph than you would at 70. If your car is air-conditioned, you'll get far better mileage with the air conditioning off than with it on. The U.S. Department of Transportation chart below graphically illustrates this:

Speed (mph)	A/C On (mpg)	A/C Off (mpg)	Saving (mpg)	Saving (%)
30	18.14	20.05	1.91	10.53
40	17.51	19.71	2.20	12.56
50	16.42	18.29	1.87	11.39
60	15.00	16.25	1.25	8.33
70	13.17	14.18	1.01	7.67

Your Home

Approximately one fifth of all our energy requirements is expended on America's 70 million homes. Of this, about 53 percent goes for heating and air conditioning, 16 percent for lighting, 15 percent for hot water, and the balance, 16 percent, for appliances.

Looking at your home from top to bottom, here are ways to save money:

INSULATION

Insulation pays its way—approximately six inches of insulation in attic spaces and three inches in the walls can save up to 30 percent on your heating bill. While it is common practice to think of insulation in terms of inches, the industry rates insulation in terms of "R" values. An "R" value unit measures an insulating material's ability to resist heat flow. The higher the "R" value, the more effective the insulation.

Here's a useful chart to show you how quickly you can save money:

CAULK AND WEATHERSTRIPPING

This is a relatively easy do-it-yourself job. Caulking and weatherstripping doors and windows can save 10 percent on your fuel bill. A gap of just one eighth of an inch on a standard-size door is the equivalent of a twenty-seven-square-inch hole in your house, leaking hot air in winter, cold air in summer.

OTHER WINDOW WISDOM

During winter, keep the curtains open on the sunny side of the house and benefit from a flow of solar heating. During summer, keep them closed and your air conditioning bill will go down an average of seven percent. Speaking of curtains and drapes, be

You Can Save Energy and Money at Home*

When You	Cost For Improvement	Annual Heating Savings	Payback Period
Upgrade ceiling insulation from zero insulation to 6" loose fill	$105.00	$128.90	About 1 Yr.
Upgrade ceiling insulation from zero insulation to 6" fiber batting	$ 99.00	$127.51	About 1 Yr.
Install storm windows and doors; caulk and weatherstrip**	$635.00	$69.58	About 9 Yrs.
Caulk and weatherstrip without adding storm windows and doors**	$ 10.00	$32.43	Less than 1 Yr.

*Based on two-story home in Washington, D.C., area (4,500 degree days) with 1,500-foot living areas.
**No caulking or weatherstripping compared with full caulking and weatherstripping.
(Source: American Gas Association)

sure that neither these, nor rugs, obscure your vents, making air flow inefficient.

COMBINATION SCREEN AND STORM DOORS

Glass is an exceptionally poor insulator. It lets heat in during the summer and out in winter. Sensible combination screen-storm windows can save you up to 15 percent on your annual fuel bill.

CHIMNEYS

Keep your chimney flue closed. Homeowners neglect to close their fireplace flues when not in use. Experts estimate that this can add 17 percent to your fuel bill over the course of a year.

ATTIC VENTILATING FAN

This will remove trapped hot air (which often can reach temperatures of up to 150°) during the summer. Thus, it markedly lowers your air-conditioning bill.

COMPARTMENTALIZE YOUR HOUSE

Seal off unnecessary rooms. Just as ships are made with bulkheads to prevent sinking, you should regard your living spaces as separate bulkheads to avoid that sinking feeling from soaring bills. Close off rooms you are not using to save on heat or air-conditioning.

FIX YOUR FILTERS

Clean filters on your heating unit and air conditioning regularly for maximum efficiency.

REGULAR MAINTENANCE

Have your heating, central-air-conditioning, and water-tank units examined at least once a year for proper maintenance.

HEATING THERMOSTATS

Set these at a maximum of 65 during the day and 55 at night. By turning your thermostat down, you really save, as this chart shows:

Temperature	Percent of Oil Saved Every 8 Hours
70°	0
65°	5
60°	9½
55°	14

Don't fiddle with the dial; this wastes heat. Don't set your thermostat higher than you ultimately want it. This doesn't cause it to heat faster.

BUYING AN AIR CONDITIONER

Manufacturers working with the U.S. Department of Commerce have started a Voluntary Energy Conservation Labeling Program. Participating manufacturers list the energy-efficiency rating (EER) on their units. Buy the most efficient one that suits your need.

OTHER AIR-CONDITIONING TIPS

Set your air conditioner at 78–80. This is the most efficient setting. During really hot days, set the fan on high. During really humid days, set the fan on low to give you more moisture. If you are going to be out for long, turn the air conditioning off.

LIGHTING

To lower energy, turn out unneeded lights. One 100-watt bulb on for an unnecessary ten hours is the equivalent of burning a pound of coal. Reducing bulb power offers major savings. Go fluorescent; it is vastly more efficient. A 40-watt standard bulb uses as much electricity as a 100-watt fluorescent bulb.

Major Electric Users

	Kilowatt Hours Per Year	Price Per Year (In Dollars)
Quick-recovery water heater	4,811	115.46
Refrigerator-Freezer		
Manual defrost (10–15 cu. ft.)	700	16.80
Automatic defrost		
(16–18 cu. ft.)	1,795	43.08
Automatic defrost		
(20 cu. ft. and up)	1,895	45.48
Freezer (15–21 cu. ft.)		
Chest type—manual defrost	1,320	31.68
Upright type		
Manual defrost	1,320	31.68
Automatic defrost	1,985	47.64
Range		
With oven	1,175	28.20
With self-cleaning oven	1,205	28.92
Clothes dryer	993	23.83
Room air conditioner	668	16.51
(Based on 800 hours of operation per year. This figure will vary widely depending on area and specific size of unit.)		
Central air conditioning	4,800	115.20
Electric resistance heating	21,000	504.00
Heat pump	14,000	336.00
Radio-record player	109	2.61
Television		
Black-and-white (solid-state)	100	2.40
Color (solid-state)	320	7.68
Attic fan	291	6.98
Dehumidifier	377	9.04
Automatic washing machine	103	2.47
Broiler	85	2.04
Coffee maker	106	2.54
Dishwasher	363	8.71
Mixer	2	0.04
Microwave oven	190	4.56
Clock	17	0.40
Hair dryer	14	0.33

Major Electric Users (cont.)

	Kilowatt Hours Per Year	Price Per Year (In Dollars)
Sewing machine	11	0.26
Shaver	0.5	0.01
Vacuum cleaner	46	1.10

HOT WATER

Insulate your hot-water tank with lagging, and insulate the pipe as well (the cold pipe, too, to prevent freezing). Check for faulty faucets. A tap leaking one drop a second will waste seven hundred gallons of hot water a year.

SHOWER, DON'T BATHE

Showers use far less water than tubs. Fine-spray nozzles use far less water than strong-spray ones.

APPLIANCES

The Edison Electric Institute has compiled national averages for appliance use. These, and the actual cost you would pay, based on 2.4 cents per kilowatt hour, are shown in the chart below:

ADDITIONAL APPLIANCE TIPS

Gang up your washing and dishwashing to save on energy. The average dish washer uses fourteen gallons to do one load—make sure it's a full one. By the same token, when you use your oven, try to cook two or more items at the same time.

Try to use your major appliances mostly early in the morning or late at night. This levels out the "energy load" for your entire community.

Defrost your refrigerator regularly.

Don't preheat your oven for more than ten minutes —it's not necessary. (Speaking of ovens, every time you take a peek at how your food is doing you waste 20 percent of the oven's efficiency.)

If you aren't listening to the radio or watching the television, turn it off.

All these are seemingly little steps—but in combination, experts estimate, they could save up to 8 percent of our total national fuel bill.

Your Lifestyle and Your Community

Each of us can, and should, save energy in other ways every day. Here are some suggestions:

1. Assist in recycling programs in your neighborhood and, when possible, use recycled products. These are invariably more energy-efficient than creating products from scratch.

2. Insofar as is possible, buy durable products. As demonstrated throughout this book, a disposable society must become a thing of the past.

3. When buying a car, a major appliance, or other energy-consuming device, take into account both the purchase price and the operating cost.

4. When buying appliances, and cars, heed the energy-consumption figures given and make every effort to buy those with the most favorable EER (energy-efficient rating).

5. Be sure you turn out all lights in your office at night. Make sure that other office equipment is switched off, such as photocopiers, duplicating machines, and electric typewriters.

6. Insofar as is possible, dispense with wasteful outdoor lighting. This can range from office parking lots after everyone has gone home to the decorative outdoor gas lamp on your front lawn (fourteen of these consume enough fuel to heat two homes).

7. In your community, see that public buildings

keep to the same heating-cooling standards you apply in your own home: in winter, 65° by day, 55° by night; in the summer, air conditioning no lower than 78°.

8. If you are a member of the local PTA, serve on your town council, or have some other public capacity, see that the overall energy cost is factored into future purchases—for example, that diesel school buses are considered or that solar heating is investigated. The same rules apply to outside organizations you may be an officer of.

For More Information

If you want to become a crusader, here are some useful addresses:

Bread for the World
235 East 49th St.
New York, N.Y. 10017

Center for Growth Alternatives
1785 Massachusetts Ave., NW
Washington, D.C. 20036

Center for International Environment Information
345 East 46th St.
New York, N.Y. 10017

Center for Science in the Public Interest
1779 Church St.
Washington, D.C. 20036

Concern Inc.
2233 Wisconsin Ave., NW
Washington, D.C. 20007

The Conservation Foundation
1717 Massachusetts Ave., NW
Washington, D.C. 20036

Department of Agriculture
14th and Jefferson Dr., SW
Washington, D.C. 20250

Energy Research & Development Administration
Washington, D.C. 20545

Energy Resources Council
Washington, D.C. 20461

Environmental Action Foundation Inc.
Room 720 DuPont Circle Building
Washington, D.C. 20036

Environmental Defense Fund
1525 18th St., NW
Washington, D.C. 20036

Environmental Protection Agency
401 M St., SW
Washington, D.C. 20460

Friends of the Earth
620 C St., SE
Washington, D.C. 20003

Institute for Food and Development
2888 Mission St.
San Francisco, Calif. 94110

International Institute for Environment & Development
27 Mortimer Street
London W1A 4QW
England

in the U.S.:
1525 New Hampshire
   Avenue NW
Washington, D.C. 20036

International Union for the
   Conservation of Nature
Morges, Switzerland

League of Conservation
   Voters
324 C St., SE
Washington, D.C. 20003

League of Women Voters
1730 M St., NW
Washington, D.C. 20036

National Audubon Society
950 Third Avenue
New York, N.Y. 10022

National Oceanic & Atmo-
   spheric Administration
Washington, D.C. 20230

National Wildlife Federa-
   tion
1412 16th St., NW
Washington, D.C. 20036

Natural Resources Defense
   Council
Washington, D.C.

Overseas Development
   Council
1717 Massachusetts Ave.,
   NW
Washington, D.C. 20036

Population Crisis Commit-
   tee
Suite 200
1835 K St., NW
Washington, D.C. 20006

The Population Institute
110 Maryland Ave., NE
Washington, D.C. 20003

President's Council on Environmental Quality
722 Jackson Place, NW
Washington, D.C. 20006

Resources for the Future
1755 Massachusetts Ave.,
   NW
Washington, D.C. 20036

Scientists Institute for Public Information
30 East 68th St.
New York, N.Y. 10021

Sierra Club
530 Bush St.
San Francisco, Calif. 94108

United Nations Environ-
   ment Program
P.O. Box 30552
Nairobi, Kenya
   New York Liaison Office
   P.O. Box 20
   Grand Central Station
   New York, N.Y. 10017

World Wildlife Fund
1319 18th St. NW
Washington, D.C. 20036

Weather Glossary

albedo. Capacity of the earth's surface to mirror the sun's heat and light back into space. In general, the lighter the color of the surface, the more solar radiation is reflected. The albedo factor is expressed in percentages. Typical albedo powers are: concrete, 17–27 percent (of the sun's energy reflected back); green forests, 5–10 percent; deserts, 25–30 percent; snow and ice, 45–90 percent; clouds, 52–85 percent (the thicker the cloud, the more it reflects back). The lowest albedo factor is the surface of water (salt and fresh). A calm ocean under noon sunlight reflects back a mere 2 percent, and absorbs the rest to help fuel the weather engine.

Antarctica. Major continent more than five times the size of Europe, which lies over and around the south pole. It is almost wholly covered by a vast ice sheet an average of sixty-five hundred feet thick. It can be roughly divided into two subcontinents: East Antarctica, the larger part, which is mainly a high plateau, and West Antarctica, mainly an icebound archipelago (chain of islands).

Antarctic Circle. The Antarctic Circle is the southern counterpart of the *Arctic Circle*. On any given date its conditions of darkness or daylight are exactly opposite to those in the north.

Arctic Circle. Line of latitude around the earth at approximately 66°30′ N. It marks the southern edge of the area within which, for one day or more each year, the sun does not set (about June 21), or rise (about December 21).

Arctic Ocean. Smallest of the world's oceans, centered approximately over the north pole. Surrounded by the landmasses of Greenland, North America, and Eurasia. Because of the surrounding land, warm ocean water cannot penetrate the Arctic Basin; the ocean around the pole stays frozen throughout the year.

blizzard. A snow storm with winds that exceed 32 mph, and snowfall dense enough to limit visibility to 500 feet (US Weather Bureau definition). A 'severe' blizzard combines near zero visibility, winds of over 45 mph and temperatures below −12°C. Antarctic blizzards are often powered by winds of well over 100 mph. Blizzard conditions can last for brief minutes, or continue for days.

carbon-dating. Method of dating remains of organic material, especially wood, from the amount of radioactive carbon (carbon 14) contained in them; developed by Nobel Prize winner Willard F. Libby and colleagues shortly after World War II. Carbon 14 occurs continuously in nature as a result of interaction between cosmic rays and nitrogen atoms, and disintegrates during the succeeding 5,730 years. The gradient of this dissolution is measurable, most accurately in the growth rings of ancient trees. Because old wood can be dated accurately by counting tree rings, studying old trees provides a means to calibrate the radio carbon calendar. With this calibration, radio carbon measurements from other remains can be used with confidence to date the time when the organisms were alive. Although it has been shown that carbon dating results do not exactly match astronomic evidence, it is a most useful method of dating samples between about eight thousand and five hundred years old.

circumpolar vortex. The swirl of the upper westerly winds around both poles.

climate. The condition of the atmosphere at a particular place on the earth over a long period of time, usually set arbitrarily as twenty to thirty years. However, it is now clear that climate is always changing and cannot strictly be regarded as "constant" even on this scale.

climatic optimum. Period of comparative warmth that began about ten thousand years ago, a period when human societies, the Mesolithic and Neolithic, for example, flourished. It ended between 5000 B.C. and 3000 B.C.

convection. The mechanism that causes the transfer of heat upward and downward. Caused by the vertical movement of heated water or air. Just as steam is given off as boiling water bubbles, so heated air rises. Air convection currents are caused by the heat of the sun, and by cold surface masses such as cold water and mountain tops. Solar heat warms the ground by radiation; this heat is given up to warm air at the ground, and that air then rises, being replaced by cool, descending air. The whole globe is involved in this mighty process of heat transference from one region to another. Convection can cause clouds and thunderstorms.

cyclonic storms. Violent revolving tropical storms that occur in both hemispheres with surface winds that exceed 75 mph. From fifty to five hundred miles in diameter, they can cause immense damage to life and property. Near the center of the whirling cloudmass is the cyclonic vortex, the "eye," bounded by a wall of spinning winds. The eye spins clockwise in the Southern Hemisphere, counterclockwise in the Northern; the wind force immediately surrounding it is often 200 mph. In the eye itself the wind drops to a light breeze; there can even be complete calm. The eye averages fifteen miles in diameter. A mature cyclonic storm several days old (they can last more than a week) can lift more than three thousand million tons of air an hour, and transport huge quantities of water over several degrees of latitude. There are between thirty and one hundred of these storms each year, about one quarter of which occur in Southeast Asia. Always spawned at sea, cyclonic storms advance at 12 to 15 mph. They are called cyclones or willie-willies in the Australian and Indian Ocean area, *typhoons* in the East Indies and China Sea region, and *hurricanes* in the North Atlantic. See also *tornado*.

doldrums. Area near the equator where the light southeast and northeast trade winds meet to produce a calm. In the days of sailing ships sailors dreaded the slack currents and fitful breezes. They are at their most pronounced along the equator in the Indian and western Pacific oceans, and slightly north of the equator off the west coast of Africa and Central America. See also *trade winds*.

easterlies. Eastward-blowing large-scale winds. Polar easterlies occur to the pole side of latitude 60°. Their tropical counterparts, the Krakatoa easterlies, blow over the equatorial region at heights between twelve and twenty-five miles.

evaporation. The rising of water vapor into the air from the heated surface of the earth's oceans, lakes, rivers, streams, and reservoirs. Warm molecules at the surface of the water, heated by the sun's radiation, are removed by wind currents moving over the water's surface. Drier air moving in absorbs more water by evaporation, and so the process goes on. Warm air absorbs water more efficiently than cold air. Like *convection,* evaporation plays a major role in the transfer of heat around the globe. Seawater—about 75 percent of the world's water—provides less evaporation than fresh water.

glacier. Land ice caused when snow falling on already compacted snow is compressed. It is estimated that more than 6 million square miles, or about 10.4 percent of the earth's land surface, are permanently glaciated. The world's longest glacier is the Lambert

Glacier; discovered in Australian Antarctic territory in the 1950s, it is up to forty miles wide, two hundred fifty miles long.

Green Revolution. New hybrid grains, fertilizers, pesticides, irrigation schemes and other technology developed in an attempt to ensure that the world always has plenty of food. Developed by agronomists like Nobel Prize winner Dr. Norman Borlaug, the new hybrid yields were at first excellent. In the 1960s global crop yields increased by 2.8 percent each year. But the droughts and floods of 1972 revealed an innate weakness—that they offer less resistance to weather changes than older, indigenous crop varieties. Today, great efforts are being made to devise new, tougher hybrids to stand up to weather extremes.

Gulf Stream. Name given to warm ocean current that flows from the Gulf of Mexico, through the Florida strait and north up the east coast of America before flowing north eastward from the U.S. coast. It plays a major role in warming the coasts and climate of Western Europe. For instance, the Norwegian port of Bergen, near which it flows, has an average temperature of 34°F for its coldest month, while Halifax, Nova Scotia, nearly one thousand miles nearer the equator but away from the Gulf Stream, averages only 23°F during its coldest month. The winter air over the Gulf Stream west of Norway is more than 40°F warmer than adjacent sea areas—the greatest temperature anomaly in the world.

horse latitudes. Two subtropical high pressure belts that girdle the earth around latitudes 30 to 35 N and S and generate light winds and clear skies. They also bring dryness to the areas below them. They are strongest over the oceans, and the Southern Hemisphere, with its greater expanse of ocean, has the larger and more continuous belt. Possibly named because becalmed sailors sometimes threw horses overboard to save on water.

THE WEATHER CONSPIRACY

hurricane. A *cyclonic storm* born in the warm waters
between the African coast and the Caribbean, usually
in late August, September, and October. The closer
to Africa they are when they grow, the more awesome
their power. "Hurricane" comes from the Caribbean
"hurucan": the devil. Regions most at risk are the
Caribbean Islands, Mexico, and the southeast corner
of the U.S. Winds of 250 mph—the highest ever
recorded—hit the Florida Keys on September 21,
1935, and killed 408. On September 8, thirty-five
years earlier, a hurricane ripped through Galveston,
Texas, and killed 6,000 people. In 1972 Hurricane
Agnes struck the coastal zone from Florida up to
Pennsylvania between June 19 and 22. It uprooted
550 bodies from a cemetery and deposited them, some
still in coffins, three miles away.

icebergs. Pieces of ice, broken off from freshwater
glaciers, that drift over the world's oceans. In previous ice ages these gigantic masses floated as far
south as the Gulf of Mexico. Even when the climate
is benign they are a considerable danger to shipping.
During this last winter they increased in number and
volume in the North Atlantic shipping lanes.

interglacial. Periods of more moderate climatic conditions between glacial advances.

jet stream. Strong, narrow air currents many miles deep
flowing at speeds of more than 100 feet per second in
the *stratosphere* from west to east. Their paths are
generally horizontal, but in some regions there are
powerful switchbacklike vertical and horizontal
changes of direction. Discovered during World War
II by bomber crews over the Mediterranean Sea and
over Japan, they play an important role in forming
high and low-pressure centers.

latitudes. Measurement of location girdling the earth
north and south of the equator. Each latitude covers
about 69 miles. Nil latitude is at the equator, and

148

these artificial impositions on the earth's surface reach north and south to 90 N and 90 S. This gives a latitude distance from pole to pole of about 12,500 miles, a diameter of 25,000 miles.

mistral. A cold dry northerly wind in southern France. It blows continuously for several days at a time, and is strongest in winter and spring. Often attains speeds of more than 60 mph and has been recorded at over 90 mph. A regular climatic feature, it can cause grave damage to crops.

monsoons. Rain winds that play an integral and perhaps major part in the world's food-climate equation. Seasonal, they are strongest along tropical coastal regions such as southern and eastern Asia. They blow offshore in winter, onshore in summer. Principal monsoons are the Indian, the Malaysian-Australian, and the West Africa. Northern Hemisphere monsoons occur between November and March, those in the Southern Hemisphere between April and October. Each year thousands of millions of people depend for their very survival on the regular arrival of the monsoons. Even a small delay can grievously reduce crop yields.

oxygen 18 dating. Method of calculating whether the weather in past millennia was warm or cold at any given time. Oxygen atoms exist in two stable forms, or isotopes. Although these are chemically identical, each atom of oxygen 18 contains two more neutral particles (neutrons) than an atom of oxygen 16, and this makes oxygen 18 heavier than the other isotope. Heavy oxygen 18 atoms evaporate less readily than oxygen 16 atoms, so the incidence of oxygen 18 levels in ice cores provides a yardstick of coldness: the less oxygen 18, the colder the climate was. Professor Willi Dansgaard and his Copenhagen University team developed this technique in the 1960s.

pack ice. Frozen seawater, with an average thickness of

149

eleven and a half feet, broken up and jammed together by wind and ocean flow. In winter, pack ice covers about 5 percent of the northern oceans, and 8 percent of the southern. The growth of the ice pack is a prime indicator of increasing cold; their shrinkage indicates warmth.

Pleistocene epoch. Epoch during which four, possibly five, vast continental ice sheets moved across the landmasses of the Northern Hemisphere. Began two and a half million years ago, ended when present *interglacial* started about ten thousand years ago.

sirocco. A warm humid wind over southern Europe and the Mediterranean which brings rain and fog. Starts in North Africa as a dry wind, then absorbs moisture over the Mediterranean Sea.

Sahel. Drought-prone part of Africa running in a wide belt below the Sahara from West Africa eastward to the Sudan. This arid corridor takes in northern Senegal, southern Mauritania, parts of Mali, Niger, Nigeria, and Chad. At least eight months of the year are dry.

stratosphere. A layer of the world's encircling atmosphere. Its lower boundary is the *troposphere,* six to eight miles above the earth's surface; its upper limit is the stratopause, about thirty miles above the earth's surface.

sunshine. The sun's visible power. Its distribution varies from four thousand hours a year in the Sahara region (about 90 percent the theoretically possible maximum daylight hours) to less than two thousand hours in stormy regions such as Ireland and Scotland.

sunspots. Swirls of gas on the sun's surface, associated with immensely powerful magnetic activity within the sun. Some of these, seen as dark blotches, may be several times the size of the earth. Sunspots can last for months; they occur in groups, pairs, and more rarely as a single phenomenon.

temperate zone. Extends from the Tropic of Cancer to the Arctic Circle in the north and from the Tropic of Capricorn to the Antarctic Circle in the south. Covers large areas of the mid-latitude westerly wind belts, and includes the U.S. Pacific Northwest coast, most of Europe, and the eastern U.S., southern Australia, and southwestern South America.

tornado. Nature's most violent and unpredictable storm, "the cat" of the cyclone family. It pounces suddenly and furiously, and causes tremendous damage in its minutes-long life. Spawned by towering thunderclouds, the tornado grows rapidly in a dirty-colored pendant cloud. A funnel spinning at a speed of between 200 and 450 mph twists from the cloud (hence the slang name "twister") and sucks up everything in its path as it touches the ground like water being sucked down a drain. Sometimes two twisters grow. The tornado forms from a severe downdraft of air from the towering clouds. Usually about 200 yards wide, the tornado may travel forward, backward, or sideways at speeds of up to 75 mph. Occasionally there are giant tornados, a mile wide and several hundred yards long, which create more havoc in the space of an hour than a major hurricane can wreak in days. Tornadoes occur principally in the American Gulf states and the Midwest, where Oklahoma and Illinois are most at risk. In all but two of the past fifty years, the U.S. has experienced more than one hundred tornados. 1973 was the worst year yet, with more than a thousand. Tornados killed an average of 230 people each year in the U.S. between 1916 and 1952.

trade winds. Steady winds that blow at an average speed of 11–13 mph from east to west, and also toward the equator from the subtropical latitudes 30 N and 30S. They produce fairly clear skies. In the days of sail, trade winds played a vital role in opening up

THE WEATHER CONSPIRACY

commercial, political, and cultural links between widely separated parts of the world.

tropics. Region of the Earth's surface immediately north and south of the equator. Bounded, approximately, by the Tropic of Cancer at 23.5 north and by the Tropic of Capricorn at latitude 23.5 south. The warmest region on earth, its seasons are marked by changes in rainfall levels; temperatures remain fairly constant throughout the year.

troposphere. Boilerhouse and to some extent nursery of our weather. Most of the clouds and weather systems are contained in this atmospheric layer, eight miles deep, sandwiched between the earth and the stratosphere. The troposphere is dominated by atmosphere convection.

tundra. Treeless bare ground in the polar regions (Arctic tundra) or on high ground (Alpine tundra). Though the climate is too cold to allow tree growth, snow cover is only intermittent. Either nothing grows except wiry grass, or the bleak landscape is characterized by low shrubs, lichens, and mosses.

turbulence. Irregular air movement when winds constantly vary in speed and direction. An important climatic factor: It churns up the atmosphere and diffuses water vapor, smoke, and dust. Turbulence at low levels reaches a maximum around midday, when the replacement of warm air by denser, colder air churns up the winds. At height, turbulence is much reduced, but may cause mighty thunderstorms. Clear-air turbulence occurs in cloudless air four to seven miles up and can be extremely violent, especially over mountain ranges. Since 1964 airliners have been equipped with a cockpit device called "cat-spy" to sniff out turbulence ten miles ahead.

typhoons. Cyclonic storms in the East Indies and along the eastern seaboard of Asia between the Philippines and Japan. The Moroto typhoon on September 21, 1934, was the most powerful to hit Japan. Its diameter was over 1,000 miles and it killed more than 3,000 people and destroyed 45,600 houses.

weather. The interaction at any given time, between the climatic variables: temperature, precipitation, wind, humidity, air pressure, cloud cover, and visibility. If it rains today, that is a facet of weather; if rainfall returns in the same month every year, like the monsoon, that is a facet of climate.

westerlies. Global belts of winds blowing from the West between latitudes 35° and 60° in both hemispheres. The wind flow, always deep, can reach a depth of nine miles or more in winter. Speeds can reach over 200 mph in the *jet stream.* Westerly winds incubate the greatest number of storms.

wind chill. The difference between air temperature and what the wind delivers on exposed bodies. Human flesh freezes solid if it is exposed to a 30 mph wind at −30°F. At 20°F a wind of 10 mph will cause a 17 degree drop in temperature.

Bibliography

All-Year Energy Savings All Around The House. Potomac Electric Power Company, Washington, D.C., 1976

Buying Solar. Federal Energy Administration. U.S. Government Printing Office, Washington, D.C., 1976

Catalogue of Publications of the World Meteorological Organization 1951–1975. Secretariat of the WMO. Geneva, 1975

Climate and the Affairs of Men. Nels Winkless III and Iben Browning. New York: Harper's Magazine Press, 1975

Climate and Health Workshop: Summary and Recommendations. U.S. Department of Commerce, U.S. Asheville, North Carolina, 1976

Climate: Present, Past and Future. H. H. Lamb. London: Methuen & Co. Ltd., 1972

Climatic Change. Harlow Shapley. Cambridge: Harvard University Press, 1953

Conservation of Energy. Shell Printing Limited, London

A Consumer Guide to Energy Conservation. American Gas Association, Arlington, Virginia

The Cooling. Lowell Ponte. Englewood Cliffs, New Jersey: Prentice-Hall, 1976

Cost and Structure of Meteorological Services with Special Reference to the Problems of Developing Countries. Secretariat of the World Meteorological Organization, Geneva, 1975

Crop Production 1976 Annual Summary. U.S. Department of Agriculture, Statistical Reporting Service, Washington, D.C., 1977

Crop Reporting Board Catalog, 1977 Releases. U.S.

Department of Agriculture, Statistical Reporting Service, Washington, D.C., 1976

Deadline Disaster. Michael Wynn Jones. Chicago: Henry Regnery Company, 1976

Degree Days. Fuel Efficiency Booklet No. 7. H. M. Stationery Office, London

Design for a Limited Planet. Norman Skurka and Jon Naar. Foreword by Jacques-Yves Cousteau. New York: Ballantine Books, 1976

Disaster Illustrated. Woody Gelman and Barbara Jackson. New York: Harmony Books, 1976

Disaster Preparedness. Report to the Congress by the Executive Office of the President, Office of Emergency Preparedness. U.S. Government Printing Office, Washington, D.C., 1972

Doing Something About the Weather. Victor Boesen. New York: G. P. Putman's Sons, 1975

Don't Be Fuelish. Federal Energy Administration. Washington, D.C., 1976

The Doomsday Book. G. Rattray Taylor. London: Thames & Hudson, 1970

Energy Conservation. Washington Gas Light Company. Washington, D.C.

The Energy Crisis and the Future. Joseph Barnea. New York: Unitar, 1975

Energy—Environment Materials Guide. Kathryn E. Mervine and Rebecca E. Crawley. National Science Teachers Association, Washington, D.C., 1975

Energy—Environment Source Book. John W. Fowler. National Science Teachers Association, Washington, D.C., 1975

Energy Labeling of Household Appliances. U.S. Department of Commerce National Bureau of Statistics. Washington, D.C.

Energy Prospects to 1985. Organization for Economic Co-operation and Development. Paris, 1974

Energy Saving in the Home. Department of Energy, London, 1976

Energy Saving in Industry. Department of Energy, London, 1975

Energy: To Use or Abuse? John David. Shell U.K. Limited, London, 1976

Forecasts, Famines, and Freezes. John Gribbin. New York: Walker and Company, 1976

The Genesis Strategy. Stephen H. Schneider, M.D., with Lynne E. Mesirow. New York and London: Plenum Press, 1976

Harvest from Weather. World Meteorological Organization, Geneva, 1967

Hearings before the Subcommittee on the Environment and the Atmosphere of the Committee on Science and Technology, U.S. House of Representatives. United States Government Printing Office, Washington, D.C., 1976

How to Conserve Gas Energy. Washington Gas Light Company, Washington, D.C.

The Hunting Hypothesis. Robert Ardrey. London: Collins, 1973

Insulate. It's Worth the Energy . . . and the Money. Potomac Electric Power Company, Washington, D.C., 1977

Live on the Cool Side. Pamphlet, Mobil Oil Corporation. 150 East Forty-second St. New York, N.Y.

National Energy Outlook. Federal Energy Administration. United States Government Printing Office, Washington, D.C., 1976

A National Plan for Energy Research, Development and Demonstration. U.S. Energy Research and Development Administration. United States Government Printing Office, Washington, D.C., 1976

Natural Gas Energy Newsletter. American Gas Association, Arlington, Virginia

The Next 200 Years. A Scenario for America and the World. Herman Kahn. New York: William Morrow and Company, 1976

Plagues and Peoples. William H. McNeill. Garden City, New York: Anchor Press/Doubleday, 1976

A Primer on Climatic Variation and Change. United States Government Printing Office, Washington, D.C.

Report of the World Food Conference, Rome, 1974. United Nations, New York, 1975

Research Aid. USSR: The Impact of Recent Climate Change on Grain Production. Central Intelligence Agency, Washington, D.C., 1976

Saving Energy Makes Cents. Washington Gas Light Company, Washington, D.C.

Speaking of Energy. Federal Energy Administration, Washington, D.C., 1975

The State of Food and Agriculture 1975. Food and Agriculture Organization of the United Nations, Rome

The Story of Natural Gas Energy. Harold W. Springlorn. Educational Services, American Gas Association, Arlington, Va., 1974

Technical Reports of the Federal Energy Administration. National Energy Information Center, Washington, D.C.

This Hungry World. Ray Vicker. New York: Charles Scribner's Sons, 1975

This Little Planet. Michael Hamilton, ed. with an introduction by Senator Edward Muskie. New York: Charles Scribner's Sons, 1970

Tips for Energy Savers. Federal Energy Administration. Washington, D.C.

Total Energy Management. Federal Energy Administration. Washington, D.C.; 1976

Toward a National Energy Policy. Mobil Oil Corporation, New York

The Unfinished Agenda. Gerald O. Barney, ed. New York: Thomas Y. Cromwell, 1977

The Value of the Weather. W. J. Maunder. London: Methuen & Co. Ltd., 1970

Warmth Kept in Keeps Heating & Costs Down. Department of the Environment, London, England

The Weather Machine. Nigel Calder. New York: Viking Press, 1974

Weather Wisdom. Albert Lee. Garden City, New York: Doubleday & Company, 1976

World Energy Outlook. Organization for Economic Co-operation and Development, Paris, 1977

Appendices

Appendix I

CIA Report: "A Study of Climatological Research as It Pertains to Intelligence Problems"

This document is a working paper prepared by the *Office of Research and Development of the Central Intelligence Agency* for its internal planning purposes. Therefore, the views and conclusions contained herein are those of the author and should not be interpreted as necessarily representing the official position, either expressed or implied, of the Central Intelligence Agency.

August 1974

Contents

Summary

The western world's leading climatologists have confirmed recent reports of a detrimental global climatic change. The stability of most nations is based upon a dependable source of food, but this stability will not be possible under the new climatic era. A forecast by the University of Wisconsin projects that the earth's climate is returning to that of the neo-boreal era (1600–1850)—an era of drought, famine, and political unrest in the western world.

A responsibility of the Intelligence Community is to assess a nation's capability and stability under varying internal or external pressures. The assessments normally include an analysis of the country's social, economic, political, and military sectors. The implied economic and political intelligence issues resulting from climatic change range far beyond the traditional concept of intelligence. The analysis of these issues is based upon two key questions:

- Can the Agency depend on climatology as a science to accurately project the future?
- What knowledge and understanding is available about world food production and can the consequences of a large climatic change be assessed?

Climate has not been a prime consideration of intelligence analysis because, until recently, it has not caused any significant perturbations to the status of major nations. This is so because during 50 of the last 60 years the Earth has, on the average, enjoyed the best agricultural climate since the eleventh century. An early twentieth century world food surplus hindered U.S. efforts to maintain and equalize farm production and incomes. Climate and its affect on world food production was considered to be only a minor factor not worth consideration in the complicated equation of country assessment. Food production, to meet the growing demands of a geometrically expanding world population, was always considered to be a question of matching technology and science to the problem.

The world is returning to the type of climate which has existed over the last 400 years. That is, the abnormal climate of agricultural-optimum is being replaced by a normal climate of the neo-boreal era.

The climate change began in 1960, but no one including the climatologists recognized it. Crop failures in the Soviet Union and India during the first part of the sixties were attributed to the natural fluctuation of the weather. India was supported by massive U.S. grain shipments that fed over 100 million people. To eat, the Soviets slaughtered their livestock, and Premier Nikita Khrushchev was quietly deposed.

Populations and the cost per hectare for technological investment grew exponentially. The world quietly ignored the warning provided by the 1964 crop failure and raced to keep ahead of a growing world population through massive investments in energy, technology, and biology. During the remainder of the 1960s, the climate change remained hidden in those back washes of the world where death through starvation and disease were already a common occurrence. The six West African countries south of the Sahara, known as the Sahel, including Mauretania, Senegal, Mali, Upper Volta, Niger, and Chad, became the first victims of the climate change. The failure of the African monsoon beginning in 1968 has driven these countries to the edge of economic and political ruin. They are now effectively wards of the United Nations and depend upon the United States for a majority of their food supply.

Later, in the 1970s one nation after another experienced the impact of the climatic change. The headlines from around the world told a story still not fully understood or one we don't want to face, such as:

- Burma (March 1973)—little rice for export due to drought
- North Korea (March 1973)—record high grain import reflected poor 1972 harvest

- Costa Rica and Honduras (1973)—worst drought in 50 years
- United States (April 1973)—"flood of the century along the Great Lakes"
- Japan (1973)—cold spell seriously damaged crops
- Pakistan (March 1973)—Islam planned import of U.S. grain to off-set crop failure due to drought
- Pakistan (August 1973)—worst flood in 20 years affected 2.8 million acres
- North Vietnam (September 1973)—important crop damaged by heavy rains
- Manila (March 1974)—millions in Asia face critical rice shortage
- Ecuador (April 1974)—shortage of rice reaching crisis proportion; political repercussions could threaten its stability
- USSR (June 1974)—poor weather threatens to reduce grain yields in the USSR
- China (June 1974)—droughts and floods
- India (June 1974)—monsoons late
- United States (July 1974)—heavy rain and droughts cause record loss to potential bumper crop.

During the last year every prominent country has launched a major new climatic forecasting program.

- USSR reorganized their climatic forecasting groups and replaced the head of the Hydrometeorological Service.
- Japan is planning to launch a major earth synchronous (U.S. manufactured) meterological satellite and has secured complete collection, processing, and analysis systems from the U.S.
- China has made major purchases of meteorological collection and analysis equipment from western industrial sources.
- India is studying the application of climatic modification to secure a more homogeneous distribution of moisture from an erratic drought/flood monsoon.
- The U.S. National Academy of Sciences is pre-

paring its recommendation for a National Climatic Research Program.

- The National Science Foundation (NSF) and the National Oceanographic and Atmospheric Agency (NOAA) have developed a National Climate Plan which will be presented to the Office of Management and Budget (OMB) for funding in FY 76.

Climate is now a critical factor. The politics of food will become the central issue of every government. On July 19, 1974, the Kiev Domestic Service reported massive rains and quoted an old proverb from Lvov Oblast, "The rains come not on the day for which we pray but only when we are making hay." The climate of the neo-boreal time period has arrived.

What Are the Intelligence Issues?

In 1972 the Intelligence Community was faced with two issues concerning climatology:

- No methodologies available to alert policymakers of adverse climatic change
- No tools to assess the economic and political impact of such a change.

In that year the Soviet Union lost a significant portion of its winter wheat crop when the snows failed to provide adequate cover and a sharp freeze destroyed the exposed vegetation. The summer moisture that normally is carried by the westerlies did not arrive in the Ukraine or the northern Oblasts (political-economic districts) of the Soviet Union. Hence, the wheat harvest was delayed, and a significant portion of the ripened crop found itself blanketed by the winter snow. The rest is economic history.

Since 1972, the grain crisis has intensified. Each year the world consumes approximately 1.2 billion metric tons. Since 1969 the storage of grain has decreased from 600 million metric tons to less than 100 million metric tons—a 30-day world supply.

With global climatic-induced agricultural failures of

the early 1970s, the stability of many governments has been seriously threatened. Many governments have gone to great lengths to hide their agricultural predicaments from other countries as well as from their own people. It has become increasingly imperative to determine whether 1972 was an isolated event or—as the climatologists predicted—a major shift in the world's climate.

The economic and political impact of a major climatic shift is almost beyond comprehension. Any nation with scientific knowledge of the atmospheric sciences will challenge this natural climatic change. The potential for international conflict due to controlled climate modification can be a reality in the 1970s. History has demonstrated that people and governments with nothing to lose have traditionally shown little regard for treaties and international conventions. Thus, any country could pursue a climate modification course highly detrimental to adjacent nations in order to ensure its own economic, political, or social survival.

In November of 1974 the United States will be participating in the World Food Conference in Rome, Italy. Some major issues of this conference will be the ability of the world to feed itself; how shortages will be met; and who will provide the needed food. Timely forecasting of climate and its impact on any nation is vital to the planning and execution of U.S. policy on social, economic, and political issues. The new climatic era brings a promise of famine and starvation to many areas of the world. The resultant unrest caused by the mass movement of peoples across borders as well as the attendant intelligence questions cannot be met with existing analytical tools. In addition, the Agency will be faced with tracing and anticipating climate modification undertaken by a country to relieve its own situation at the detriment of the United States. The implication of such a modification must be carefully assessed.

Climatic Phenomena and the State of the Art

Since the late 1960s, a number of foreboding climatic predictions has appeared in various climatic, meteorological, and geological periodicals, consistently following one of two themes.

- A global climatic change was underway
- This climatic change would create worldwide agriculture failures in the 1970s.

Most meteorologists argued that they could not find any justification for these predictions. The climatologists who argued for the proposition could not provide definitive causal explanations for their hypothesis.

Early in the 1970s a series of adverse climatic anomalies occurred:

- The world's snow and ice cover had increased by at least 10 to 15 percent.
- In the eastern Canadian area of the Arctic Greenland, below normal temperatures were recorded for 19 consecutive months. Nothing like this had happened in the last 100 years.
- The Moscow region suffered its worst drought in three to five hundred years.
- Drought occurred in Central America, the sub-Sahara, South Asia, China, and Australia.
- Massive floods took place in the midwestern United States.

Within a single year, adversity had visited almost every nation on the globe.

The archaeologists and the climatologists document a rather grim history of the cultural pressures instigated by changes in climatic regime. Recently, some archaeologists and historians have been revising old theories about the fall of numerous elaborate and powerful civilizations of the past, such as the Indus, the Hittites, the Mycenaean, and the Mali empire of Africa. There is considerable evidence that these empires may have been undone not by barbarian invaders but by climatic change. Bryn Mawr archaeologist, Rhys Carpenter, has tied several of these declines to specific global cool periods, major and minor, that affected the global atmospheric circulation and brought wave upon wave of drought to formerly rich agricultural lands.

Refugees from these collapsing civilizations were often able to migrate to better lands. Reid Bryson, of the University of Wisconsin's Institute for Environmental Studies, speculates that a new rainfall pattern might actually revive agriculture in some once-flourishing regions such as the northern Sahara and the Iranian plateau where Darius' armies fed. This would be of little comfort, however, to people afflicted by the southward encroachment of the Sahara. The world is too densely populated and politically divided to accommodate mass migrations.

Yet to understand the results of climatic change, we must know something of the basics of climatology and the people associated with this science.

THE STATE OF THE ART OF CLIMATOLOGY

The climate of a region on the Earth is said to be represented by a statistical collection of its weather conditions during a specified time interval. This interval is usually at least two or three decades; for the Agency's purposes, we will be dealing with months and years. Climatology, as derived from the Greek word *clima* meaning inclination of the sun's rays, reflects the importance attributed by the early students of climatology to the influence of the sun. For some unknown reason, this importance was virtually ignored by climatologists until a decade ago.

The keys to understanding climatology are:
- The acceptance of the principal that nature abhors a heterogeneous *distribution of energy*.
- *Energy reaching the Earth* is modulated by variations in the Earth's orbit, the inclination of the Earth's axis while in orbit, the materials in the

Earth's atmosphere (dust, moisture, etc.), and energy fluctuations in the sun itself.
- The *Earth's atmosphere* absorbs only a small percentage of the energy coming directly from the sun.

Thermal Distribution of Energy

Parameters such as temperature, rainfall, and wind velocity can be directly related to the heterogeneous distribution of thermal energy on the surface of the Earth. They are the physical manifestations of a global system which attempts to attain a thermal equilibrium through the interchange of potential and kinetic energy between the atmosphere and the oceans.

Energy Reaching the Earth

There are two major reasons for the heterogeneous distribution of energy on the surface of the Earth:
- Cloud formations
- Surface albedo—ratio of energy reflected to energy received from the sun.

Each affects the amount of solar energy absorbed by the Earth.

The energy required for the physical processes taking place in the Earth-atmosphere-ocean system is almost totally provided by the sun. Each minute the sun radiates approximately 56×10^{26} calories of energy of which 2×10^3 calories per square centimeter per minute are incident upon our outer atmosphere. The exact amount of solar radiation that actually is incident depends upon the time of year, the time of day, and the latitude. Bare Earth absorbs and transforms approximately 75 percent of all visible light impinging upon it. The remainder of the energy is reflected back into space.

Figure 1 depicts the atmospheric energy as it is received from the sun. The surface of the Earth, heated by the absorption of visible or short-wave solar radiation, converts this energy to thermal or long-wave radiation which, in turn, convectively and conductively heats the atmosphere.

In a typical region of scattered clouds 2 percent of the incident visible solar energy is absorbed in the stratosphere. Fifteen percent of the remaining energy is typically absorbed in the troposphere and converted to thermal energy.

Forty-seven percent of the visual radiation eventually reaches the surface—31 percent directly and 16 percent through atmospheric diffusion. Note that 36 percent of the original energy is reflected back into space—23 percent from the tops of clouds, 6 percent through diffusion in the troposphere, and 7 percent from the surface.

It is not obvious how the Earth maintains its energy balance. The Earth's surface absorbs about 124 kilolangleys of solar radiation each year. Energy per unit area is expressed in langleys (ly) or kilo-langleys (kly). One langley is equivalent to 1 calorie per square centimeter. The Earth effectively radiates 52 kly of longwave energy to the atmosphere. The difference between incoming and outgoing radiation is 72 kly which is the net energy balance. The global radiation balance is zero averaged (approximately) over the year, but it will not equal zero either seasonally or annually in a given latitude zone.

The atmosphere is uniformly a radiative heat sink at all latitudes, while the Earth's surface—except near the poles—is a heat source. Energy must therefore be transferred from the surface to the atmosphere to keep the surface from warming and the atmosphere from cooling. The vertical heat exchange occurs mainly by evaporation of water from the surface (heat loss) and condensation in the atmosphere (heat gain) and by conduction of sensible heat from the surface and turbulent diffusion into the atmosphere (convection).

An example of energy balance is represented in Figure 1. The atmosphere can gain energy from a variety of sources. The troposphere gains 15 percent of its thermal energy through direct conversion of visual energy to thermal energy. Of the black body (thermal) radiation from the Earth's surface (98 percent), 91 percent is partially absorbed in the atmosphere, and

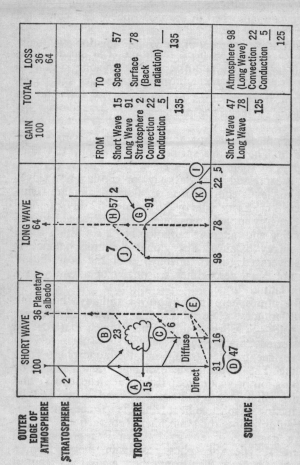

Figure 1. Atmospheric energy (in percent).

the remaining 7 percent is radiated into space. The stratosphere provides an additional 2 percent to the troposphere; convection (22 percent) and conduction (5 percent) account for about 27 percent.

The surface, then, has two sources of energy. It gains 47 percent in visual-to-thermal energy transformation and 78 percent in back radiation. The surface loses 98 percent to the atmosphere through long (infrared) waves, and 22 percent through convection and 5 percent through conduction. The gains and losses in the atmosphere-surface system are time dependent.

A layer of clouds, snow and ice can reflect 80 to 90 percent of the visible light back into space. Because climate depends primarily upon the amount of solar radiation that is absorbed by the Earth and atmosphere, albedo becomes important. Albedo is the ratio of the energy received from the sun and reflected by the Earth. The greater the albedo, the colder the Earth.

Clouds can serve to moderate whatever climate trend is under way: if the Earth's surface temperature climbs for whatever reason, more water evaporates and may rise to form more cloud cover. This increases the albedo and lowers the rate of heating. Ice and snow, on the other hand, provide positive feedback; if the average year-round temperature decreases, the extent of ice and snow coverage increases and reflects more of the incoming sunlight back to space. The result is to lower the rate of heating still more, particularly in the regions closest to the poles.

There is yet another contributor to the planet's albedo—airborne particles, particularly the extremely fine dust particles that have been carried too high in the atmosphere to be washed out by precipitation. Many of these particles remain aloft for months or years. Thus, a heterogeneous distribution of clouds may eventually cause a heterogeneous distribution of thermal energy around the Earth.

Earth's Atmosphere

Many mechanisms are employed by the Earth to bring itself into thermal equilibrium. When thermal radiation from land surfaces heats the air directly above it, the rising air causes a change in the local atmospheric pressure. The spinning of the Earth and the resultant pressure differentials are physically manifested by the gaseous currents known as "wind." Thermal radiation from the Earth also causes the evaporation-condensation cycle that forces moisture from land and ocean sources to enter the atmosphere (Figure 2). Thus, the atmosphere becomes one of the major means of equalizing the thermal energy distribution around the world.

For evaporation to occur, both a driving force and a source of energy in the transformation phase are required. Radiation is the main energy source. In the presence of an adequate supply of energy, most precipitation evaporates before it has a chance to run off. The oceans lose more water by evaporation (84 percent) than they gain by precipitation (77 percent). The deficit is made up by run off from the continents (7 percent) over which precipitation exceeds evaporation.

The oceans provide about 84 percent of global evaporation, while the continents provide the remaining 16 percent. The change of phase from a liquid state to a vapor requires that energy be provided to overcome the intermolecular attractions between water molecules. The latent heat required to evaporate one gram of water at 0°C. is 600 calories. Condensation is responsible for releasing this energy. Thus the 7 percent horizontal advection of water vapor to the land mass contributes significantly to the transfer of energy to the continents. The normal dynamic movement of air and vapor masses is continuously directed at the equalization of energy in the oceans as well as land masses.

Since 80 percent of the Earth's surface is water, principally the oceans, it would seem reasonable that mechanisms had to exist in the oceans to offset the heterogeneous distribution of thermal energy. Throughout all the major oceans, great currents of water flow

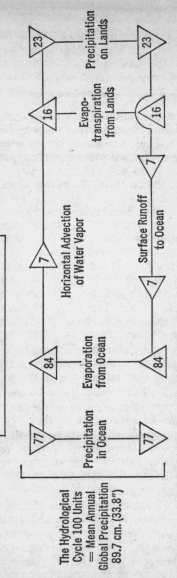

Figure 2. Atmospheric moisture (in percent).

between energy sinks and sumps. The winds created by the ocean's radiated energy form atmospheric tides. As an example, the greatest mass of water on the Earth— the Pacific Ocean—is constantly depressed one meter on the east side as compared with the west due to atmospheric pressure anomaly.

The most dangerous effect of the global cooling trend has been a change in atmospheric circulation and rainfall. The change centers on the behavior of the circumpolar vortex, the great cap of high-altitude winds revolving about the poles from west to east. The broad band across the Northern Hemisphere marks the approximate southern edge of the wind system as it was during the summertime in the early 1960s. Its southern edge determines the location of the prominent high-pressure regions, indicated here by narrow clockwise-spiraling arrows representing winds flowing outward. The highs result from dry winds that descend after traveling at high altitudes from the equator. They created the world's great deserts and determine the northern limit of penetration by rain-bearing summer monsoons. The limit is known as the "intertropical convergence zone."

Because of the global cooling trend, the lower edge of the circumpolar vortex has in recent years stayed farther south during the summer, in the position shown by the smaller band near the equator. It has kept the high pressure zones farther south too, blocking the monsoons out of regions where they are vital to the survival of hundreds of millions of people. At the same time, the vortex's semistationary wave patterns have altered, affecting rainfall patterns in temperate regions and making the climate more variable. The deeper wave over the U.S., for example, is believed responsible for recent cold winters in the West and mild ones in the East. The West has been subjected to north winds; the East, the return flow. Although some evidence exists that the cooling trend has affected wind patterns in the Southern Hemisphere as well, weather statistics are scanty.

CURRENT APPROACHES TO CLIMATOLOGY

There are three basic schools or philosophies of climatology. The first is centered around Professor H. H. Lamb, who is currently the Director of the Climatic Research Unit at the University of East Anglia in the United Kingdom. This school contends that if a climatologist is to project future climates, he must understand what has occurred in the past. The second is characterized by Dr. Joseph Smagorinsky, who is the Director of the Geophysical Fluid Dynamics Laboratory at Princeton University. This center believes that a complete understanding of atmospheric circulation is sufficient for climatic forecasting. The third is best represented by Dr. M. I. Budyko, an eminent Soviet climatological theoretician. He pursues the hypothesis that an understanding of the total distribution of thermal energy is necessary for climatic forecasting.

The Lambian school is based on the establishment of climatic statistical trends. A great deal of effort has been expended by the followers of this philosophy in quantifying the qualitative descriptions provided by historical sources (ancient court scribes, ship's logs, and scholars). Their reconstruction of climatic conditions has reached back 5,000 years. This particular approach was almost totally dependent upon man-made records. In recent years, the use of geophysical indicators such as tree rings, sedimentary deposits, and Arctic ice layering has added substantially to the global data pool.

Unique scientific methods have been developed which allow the climatologist to determine the historical intensity and distribution of solar radiation and precipitation on a worldwide basis. Before these developments, it was necessary for the scientist to infer climatic variability based on many indirect factors. Though this work is still quite incomplete, preliminary reports by Dr. John Imbrie (Brown University) have provided a fascinating portrayal of the Earth's climatology over the last 50 million years.

Current evidence indicates that the continental areas which were once in the tropical climatic regions of this planet, for some reason, underwent a rapid climatic change. Beginning approximately 20 million years ago (miocene period), large climatic variations characterized by what is known as the Ice Ages began to make appearances. Dr. Imbrie's group has been able to establish that these Ice Ages are cyclic in nature and consist of approximately a 90,000-year glacial period followed by a relatively brief warming peak for 10,000 to 12,500 years, called the interglacial periods. Thus, as we see in Figure 3 based upon a sample space of 20 million years, this rather narrow period of the interglacial span is a consistent feature.

Investigations indicate interglacial periods never extended beyond 12,500 years nor has the period ever been less than 10,000 years (Figure 4). The glacial periods may be characterized by large continental ice sheets that extended across vast regions of Europe, North America, and Asia. This phenomena is well documented on the North American continent and came to an end approximately 10,000 years ago. The present interglacial era is characterized by a thermal maximum which occurred about 5,000 to 3,000 B.C. During this time, many major deserts in the world—as we know them—were formed, such as the Sahara, the Arabian, and great Mongolian deserts.

Climate change at the end of these interglacial time periods is rather sharp and dramatic. Excellent historical evidence exists from areas on the European plains which once were oak forests and were later transformed into poplar, then into birch, and finally into tundra within a 100-year span. Thus, the researchers of the CLIMAP group (CLImatic MAPing) hypothesize that the change from an interglacial to glacial time period could take place in less than 200 years. An example of rapid climatic changes are the remains of frozen mastodons completely preserved in Siberian and North American ice packs.

Scientists are confident that unless man is able to

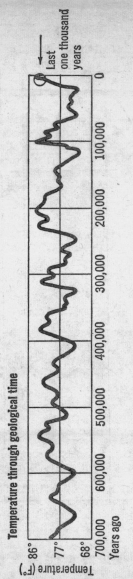

Figure 3. Temperature history through geological time.

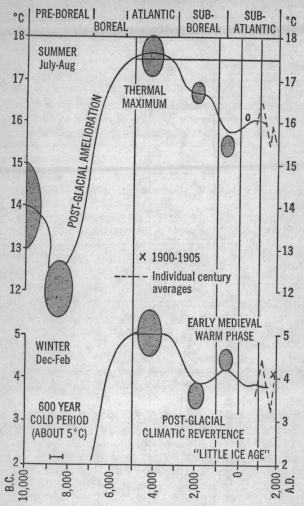

Figure 4. Climate variation in England.

effectively modify the climate, the northern regions, such as Canada, the European part of the Soviet Union, and major areas in northern China, will again be covered with 100 to 200 feet of ice and snow. That this will occur within the next 2,500 years they are quite positive; that it may occur sooner is open to speculation.

The Smagorinsky-ian school of climatology is based upon the meteorologist's attempts to extend the predictive capabilities of the equations of fluid motion. Meteorology deals principally with the forecasting of atmospheric pressure differentials and the propensity for given patterns to result in rain, snow, ice, high winds, etc. It does not take into account solar or Earth radiation nor hydrological (i.e., evaporation) variables.

Since the availability of serial, numerical computers in the latter half of the 1940s, the meteorologist has developed a system of models to predict near-term atmospheric variations. The basic tool employed by this group is the General Circulation Model. These models describe the effects of large-scale atmospheric motion and are treated explicitly by numerical integration. For almost 30 years the meteorologist has tried unsuccessfully to extend his predictive capability past a 24-hour forecast. The Smagorinsky-ian approach, however, is the currently accepted methodology within the United States Government and receives more than 90 percent of all the research and development funding available therein.

The Budyko-ian school is based upon the theoretical work of Dr. M. I. Budyko, who is associated with the Global Meterological Institute in Leningrad. The basis of this approach to the climatological problem is Dr. Budyko's 1955 paper entitled, "The Heat Balance of the Earth's Surface." This paper advances the hypothesis that all atmospheric motions are dependent upon the thermodynamic effect of a nonhomogeneous distribution of energy on the Earth's surface. Though this work originally met with opposition from the world's meteorologists, it is now accepted as a more reasonable basis for developing a successful climatic prediction model. The earlier, simplistic explanation of climate was basically Budyko-ian.

Recent Milestones

Explanation of the scientific phenomena and elaboration of the three methodological schools provide a background for more recent developments—developments which have more relevancy to requirements as they might emerge in the Intelligence Community. The University of Wisconsin's work appears to be providing the cohesion for continuing research in this area.

THE WISCONSIN STUDY

The University of Wisconsin was the first accredited academic center to forecast that a major global climatic change was underway. Their analysis of the Icelandic temperature data, which they contend has historically been a bellwether for northern hemisphere climatic conditions, indicated that the world was returning to the type of climate which prevailed during the first part of the last century. This climatic change could have far-reaching economic and social impact. They observed that the climate we have enjoyed in recent decades was extremely favorable for agriculture. During this period, from 1930 to 1960, the world population doubled, national boundaries were redrawn, the industrial revolution became a worldwide phenomenon, marginal lands began to be used in an effort to feed a vastly increased population, and special crop strains optimally suited to prevailing weather conditions were developed and became part of what was called the "green revolution."

The climate of the 1800s was far less favorable for

agriculture in most areas of the world. In the United States during that century, the midwest grain-producing areas were cooler and wetter, and snow lines of the Russian steppes lasted for longer periods of time. More extended periods of drought were noted in the areas of the Soviet Union now known as the new lands. Moreover, extensive monsoon failures were common around the world, affecting in particular China, the Philippines, and the Indian subcontinent.

The Wisconsin analysis questioned whether a return to these climatic conditions could support a population that has grown from 1.1 billion in 1850 to 3.75 billion in 1970. The Wisconsin group predicted that the climate could not support the world's population since technology offers no immediate solution. Further, world grain reserves currently amount to less than one month; thus, any delay in availability of supplies implies mass starvations. They also contended that new crop strains could not be developed overnight, and marginal lands would be less suited or perhaps unsuited to agricultural production. Moreover, they observed that agriculture would become even more energy dependent in a world of declining resources. Their "Food for Thought" chart (Figure 5) conveys some idea of the enormity of the problem and the precarious state in which most of the world's nations could find themselves *if the Wisconsin forecast is correct.*

As an example, Europe presently, with an annual mean temperature of 12°C. (about 53°F.), supports three persons per arable hectare. If, however, the temperature declines 1°C. only a little over two persons per hectare could be supported and more than 20 percent of the population could not be fed from domestic sources. China now supports over seven persons per arable hectare; a shift of 1°C. would mean it could only support four persons per hectare—a drop of over 43 percent.

A unique aspect of the Wisconsin analysis was their estimate of the duration of this climatic change. An

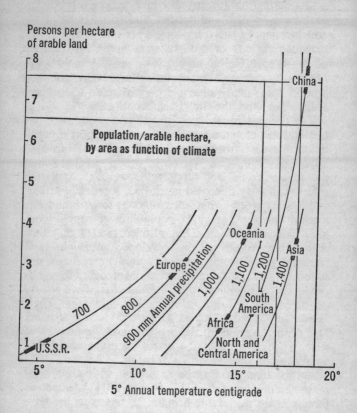

Figure 5. Food for thought.

analysis by Dr. J. E. Kutzbach (Wisconsin) on the rate of climatic changes during the preceding 1600 years indicates an ominous consistency in the rate of which the change takes place. The maximum temperature drop normally occurred within 40 years of inception. The earliest return occurred within 70 years. (Figure 6). The longest period noted was 180 years.

The study of the impact of climatic change on past and present cultures has been a cooperative venture between the social scientist, the historian, and the climatologist. It has been shown that over the last

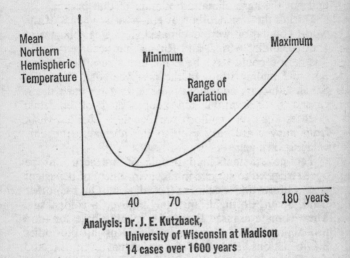

Figure 6. Mean temperature variation during a new climatic era.

10,000 years there have been many climatic changes of regional and global significance. Detailed descriptions exist showing how these climatic changes affected the people of these regions. The Wisconsin forecast suggests that the world is returning to the climatic regime that existed from the 1600s to the 1850s, normally, called the neo-boreal or "Little Ice Age." (This climate was physically characterized by broad strips of excess and deficit rainfall in the middle latitudes and extensive failure of the monsoons.) The political, historical, and economic consequences of this climatic era have heretofore been masked by the historian's preoccupation with the technical progress. We have recent evidence of this type of faulty analysis which has persuaded the modern agroeconomist that man's agricultural growth during the last 40 years was only due to technology and not the agro-climatic optimum of that period.

During the last neo-boreal era great segments of the world population were decimated. The great plagues of Europe, India, Africa, and Russia that occurred during this period could have been the direct result of starvation and malnutrition. In the past year data from the Sahel, Ethiopia, and India indicate that for each death caused by starvation, ten people died of epidemic diseases such as smallpox and cholera. Bodies weak from hunger are easy prey to the normal pathogenic enemies of man.

The governments and people of northern Europe once struggled to survive in an environment of persistent crop failure and declining population. On the other hand, Spain, Portugal, and Italy enjoyed a golden age. Their climate assured them of a reliable base for food production. The German states, Russia, the other Slavic nations, and to a certain extent even England and France, lived in the twilight of permanent winter.

For 250 years most of the world suffered major economic and political unrest which could be directly or indirectly attributed to the climate of the neo-boreal era. The great potato famine of 1845 in Ireland was the last gasp of the "little ice age." Yet for every death in Ireland there were ten in the Asian countries.

What would a return to this climate mean today? Based on the Wisconsin study, it would mean that India will have a major drought every four years and could only support three-fourths of her present population. The world reserve would have to supply 30 to 50 million metric tons of grain each year to prevent the deaths of 150 million Indians. China, with a major famine every five years, would require a supply of 50 million metric tons of grain. The Soviet Union would lose Kazakhstan for grain production thereby showing a yearly loss of 48 million metric tons of grain. Canada, a major exporter, would lose over 50 percent in production capability and 75 percent in exporting. Northern Europe would lose 25 to 30 percent of its present product capability while the Common Market countries would zero their exports.

PEOPLE, PLACES AND APPROACHES

A limited number of people within the United States are involved in climatological research. On the West Coast there are two significant groups. The first is under Dr. Larry Gates at the RAND Corporation in Santa Monica. Dr. Gates' work has been supported by ARPA and is theoretically Smagorinsky-ian. He has worked for three years under an ARPA grant utilizing basically the UCLA two-level General Circulation Model. Though the work has been theoretically interesting and has developed many new software capabilities, they have still not arrived at an operational system. Dr. Gates has been strongly impressed by developments in the Budyko-ian school and is in the process of modifying their simulation programs to incorporate some of the more recent thermodynamic developments.

The Scripps Institution group at La Jolla, under the direction of Dr. John Isaacs, and more recently with the inclusion of Dr. Jerome Namais, has followed both the Lambian and Budyko-ian approaches to climatological problems. Their main capabilities have been in the development of climatological observables. Dr. Isaacs' early work, which has been continued by Namais' research, was directed at the thermodynamic influence of the oceans on world atmospheric circulation. At present, no pragmatic climatological forecasting is being pursued at Scripps.

The atmospheric sciences group at the University of Arizona is solidly Budyko-ian. Dr. William Sellers who heads this group is one of the country's leading technicians in Budyko-ian methodology. His first published climatic model in 1968 was not well received by the Smagorinsky-ians or by the Budyko-ians within the world community. They did acknowledge, however, that it was the first pragmatic systematizing of this approach. His latest model, developed in 1972, has had a significant effect in crystalizing this whole philosophy and demonstrating a pragmatic climatological model.

There are two climate groups in the midwest—one being NCAR (National Center for Atmospheric Research) at Boulder, Colorado. Their efforts have been to explore highly disaggregated atmospheric models. The second group, at the University of Wisconsin, is under Reid Bryson and John Kutzbach, both mentioned earlier. Their work at Wisconsin represents the focal point for climatological research in the United States. They are the only people within the academic community in the United States that have a seasonal climatological forecasting system.

The eastern establishment, consisting of Princeton and the Massachusetts Institute of Technology, is primarily Smagorinsky-ian. They are basically NOAA-funded and, though primarily engaged in increasing the accuracy of meteorological forecasts, have attempted without success to provide climatological forecasting capabilities.

In summation, the eastern schools have employed basically the Smagorinsky-ian principles in one way or another. The limitation of this approach, although not yet apparent to the establishment, is rapidly being abandoned by the academic community. The pragmatic capabilities of the Budyko-ians and the methodologies therein are quickly being absorbed by both the East and West Coast establishments. The Lambians and their primarily statistical approach are beginning to lose favor, but their development of historical climatological records has provided a vital service within the climatological community.

SAN DIEGO CONFERENCE

By the fall of 1973 the Office of Research and Development (ORD) had obtained sufficient evidence to alert the Agency analysts that forecasts of an ongoing global climate change were reasonable and worthy of attention. ORD also determined that it was feasible to begin the development of forecasting techniques and impact assessment. However, Agency analysts remained skeptical, noting that the mix of approaches (Wisconsin, Scripps, RAND, NCAR) and the scientific personalities pursuing them prevented a clear expression of what the recognized authorities were agreeing on.

To resolve these issues, the principal investigators representing the various research approaches convened in San Diego in April 1974 to discuss these three specific topics:

- The state of climatological forecasting; identification of elements of the methodology wherein there is some consensus, current trends in development, and new approaches.
- Prospects for developing near-term applications of climatology to Agency interests.
- Recommendations for high- and low-risk approaches for long-range climatological models development.

For two days they argued, discussed, and defended their approaches to climatic forecasting and the impact of climatic change. By the second day a consensus was reached on the following fundamental issues:

- A global climatic change is taking place.
- We will not soon return to the climate patterns of the recent past.
- For the future, there is a high probability of increased variability in a number of features of climate that are of importance to crop growth.
- The most promising long-range (1–5 years) approach to climate forecasting appears to be the statistical synoptic approach. The consensus ex-

pressed caution in using these projections without an attempt to develop some physical understanding of the underlying weather-forcing mechanisms.

In general, the conference participants were skeptical of the prospects of making a one- to five-year forecast at this time, stating that only season-to-season forecasts were within the state of the art.

The conference participants unanimously recommended that the clear need for a long-range prediction dictated the establishment of an Operational Diagnostic Center charged with developing global forecasting techniques and for servicing the Government's needs for one-to-five year forecasts.

NATIONAL CLIMATE PLAN

In the summer of 1973 the Wisconsin Plan for Climatic Research was presented to the National Security Council. NOAA and the National Science Foundation were requested to review this plan and to suggest how it should be implemented. The Wisconsin Plan stimulated activity in many agencies.

In the fall of 1973 three agencies in the government became active in the development of climatic research plans: NSF, NOAA, and the National Academy of Sciences. The National Academy of Sciences established the Committee on Climatic Variation, chaired by Dr. Larry Gates. The committee members completed their recommendations for a National Climatic Research Plan in June of 1974. This plan is presently under assessment by the National Academy of Sciences. Its final approval is expected late this year. Early in 1974, NOAA began developing a plan which would include a Center for Climatic and Environmental Assessment as suggested by preliminary recommendations from the National Academy of Sciences Committee. This plan would allow NOAA to respond rapidly to the needs of government agencies that are concerned with the impact of climatic factors on both a national and global scale.

In the spring of 1974, the Director of the Polar

Studies Division of NSF developed a plan to establish a Center for Climatic Research as well as to provide funding to appropriate academic centers.

Both of these plans have been incorporated into what is now called the National Climate Plan. NOAA would be responsible for developing methods for practical climate forecasting as well as developing techniques applicable for the assessment of national and international food production. NSF would provide support to responsible academic centers and establish a Center for Climatic Research. This Center would operate in a similar manner as the present National Center for Atmospheric Research (NCAR) at Boulder, Colorado. The National Climatic Plan is presently under review by NOAA and the NSF. They expect to seek approval from the Office of Management and Budget in the fall of 1974 for FY 76 program funding.

Conclusions

Leaders in climatology and economics are in agreement that a climatic change is taking place and that it has already caused major economic problems throughout the world. As it becomes more apparent to the nations around the world that the current trend is indeed a long-term reality, new alignments will be made among nations to insure a secure supply of food resources. Assessing the impact of climatic change on major nations will, in the future, occupy a major portion of the Intelligence Community's assets.

Climatology is a budding science that has only recently given promise of fruition. Classical climatology was occupied with the archiving of evidence. Until 1968 very little was accomplished in this science toward defining casual relationships. During the last two years climatologists have made substantial progress in the development of methodologies and techniques in forecasting climatic changes. Recent developments in climatology have shown extensive promise toward providing seasonal forecasting. In the near future it may be possible to provide forecasts in the realm of one to five years.

The function of research within the Agency has been directed at defining the relationship of climatology to the intelligence problems. It is increasingly evident that the Intelligence Community must understand the magnitude of international threats which occur as a function of climatic change. These methodologies are necessary to forewarn us of the economic and political collapse of nations caused by a worldwide failure in food production. In addition, methodologies are also necessary to project and assess a nation's propensity to initiate militarily large-scale migrations of their people as. has been the case for the last 4,000 years.

Though the issues are important, the United States has a limited capability in climatic forecasting. The government expends over $150 million annually on short-range weather forecasting, but only a minimum of direct dollars on climatic forecasting. Only a few academic centers in the United States are engaged in training personnel in this field, which suggests we have a limited chance of solving the Intelligence Community's problem unless decisive action is taken.

Bibliography

Adem, J., 1964: "On the Physical Basis for the Numerical Prediction of Monthly and Seasonal Temperatures in the Troposphere-Ocean-Continent System," *Mon. Wea. Rev.* 92: 91–103.

Alexander, Tom, 1974: "Ominous Changes in the World's Weather," *Fortune,* p. 90.

Asakura, Tadashi, 1974: "Unusual Weather and Environmental Pollution," U.S. Joint Publications Research Service. JPRS L/4913, 22 May.

Bauer, K. G., 1971: "Linear Prediction of a Multivariate Time Series Applied to Atmospheric Scalar

Field," Ph.D. Thesis, University of Wisconsin-Madison (Unpublished), 180 pp.

Bryson, R. A. and J. A. Dutton, 1961: "Some Aspects of the Variance Spectra of Tree Rings and Varves," *Annal N. Y. Acad. Sci.* 95(1): 580-604.

Bryson, R. A., 1972: Climatic Modification by Air Pollution in *The Environmental Future,* N. Poluin (ed.). London: Macmillan, xiv + 660 (pp. 134–174).

Bryson, R. A. and J. E. Kutzbach, 1973: "On the Analysis of Pollen-Climatic Canonical Transfer Functions" (in preparation).

Bryson, R. A. and W. P. Lowry, 1955: "Synoptic Climatology of the Arizona Summer Precipitation Singularity," *Bull. Ameri. Met. Soc.* 36: 329-399.

Budyko, M. I., 1969: "The Effect of Solar Radiation Variation on the Climate of the Earth," *Tellus* 21: 611-619.

Chang, Jen-Hu, 1970: "Potential Photosynthesis and Crop Productivity," *Ann. Assoc. Am. Geogr.* 70: 92-101.

Cottam, G., E. Howell, F. Stearns, and N. Korbiger, 1972: "Productivity Profile of Wisconsin, Part II," *Eastern Deciduous Forest Memo Report* #72-142.

Davis, N. E., 1972: "The Variability of the Onset of Spring in Britain," *Quart. J. Roy. Met. Soc.* 98: 763-777.

Davitaya, F. F., 1965: "The Possible Influence of Atmospheric Dustiness on the Recession of Glaciers and Warming of the Climate," *Izvestiya Akad. Nauk SSSR Geogr. Ser.* No. 2, Mar-Apr.: 3-33.

Dwyer, H. A. and T. Peterson, 1973: "Time-dependent Global Energy Modeling," *Jour. Appl. Meteor.* 12(1): 36-42.

Fritts, H. C., T. J. Blasing, B. P. Hayden, and J. E. Kutzbach, 1971: "Multivariate Techniques for Specifying Tree-Growth and Climate Relationships and for Reconstructing Anomalies of Paleoclimate," *Jour. Appl. Meteor.* 10(5): 845-864.

Fultz, D., 1961: "Developments in Controlled Experiments on Larger Scale Geophysical Problems," *Adv. Geophys.* 7:1-103.

Ilesanmi, O. O., 1971: "An Empirical Formulation of an ITD Rainfall Model for the Tropics: A Case Study for Nigeria," *Jour. Appl. Met.* 10(5): 882-891.

Kukla, George J., and J. Helena, 1974: "Increased Surface Albedo in the Northern Hemisphere," *Science* 183(4126): 709.

Kutzbach, J. E., 1970: "Large-Scale Features of Monthly Mean Northern Hemisphere Anomaly Maps of Sea Level Pressure," *Mon. Wea. Rev.* 98(9): 708-716.

Lamb, H. H., 1966: "Climate in the 1960's," *Geographic Journal* 132: 183-212.

Lamb, H. H., 1970: "Volcanic Dust in the Atmosphere," *Phil. Trans. Roy. Soc. London* 266(1178): 425-533.

Landsberg, H., 1967: "Climate, Man, and Some World Problems," *Scientia* May-June.

Lettau, H. H. and K. Lettau, 1969: "Shortwave Radiation Climatonomy," *Tellus* 21: 208-222.

Lettau, K., 1973: "Modeling of the Annual Cycle of Soil Moisture," *Proc. of the Symposium of Phenology and Seasonality Modeling*, Springer-Verlag: New York (accepted for publication in 1973).

Leith, H., 1972: "Modeling and Primary Productivity of the World," *Nature and Resources* 8: 2-10.

Lorenz, E. N., 1970: "Climatic Change as a Mathematical Problem," *J. Appl. Met.*, 9.

Machta, Lester, 1972: "Mauna Loa and Global Trends in Air Quality," *Bull. Am. Met. Soc.* 53(5): 402-421.

McQuigg, J., S. Johnson, and J. Tudor, 1972: "Meteorological Diversity-Load Diversity, A Fresh Look at an Old Problem," *J. Appl. Met.* (11)4: 561-566.

Miller, P. C., 1971: "Bioclimate, Leaf Temperature, and Primary Production in Red Mangrove Canopies in South Florida," *Ecol. 53: 22-45.*

Mitchell, J. M. Jr., 1961: "Recent Secular Changes of

Global Temperature," *Annals of New York Academy of Sciences,* Article 1. 95: 235-250.

Monteith, J. L., 1965: "Light Distribution and Photosynthesis in Field Crops," *Ann. Bot.* 29: 17-37.

Paddock, W. and P. Paddock, 1967: *Famine—1975!* Boston: Little, Brown, & Co., x + 276 pp.

Reitan, C. H., 1971: "An Assessment of the Role of Volcanic Dust in Determining Modern Changes in the Temperature of the Northern Hemisphere," Ph.D. Thesis, University of Wisconsin-Madison (unpublished), 147 pp.

Sellers, W. D., 1969: "A Global Climatic Model Based on the Energy Balance of the Earth Atmosphere System," *J. App. Met.* 8: 392-400.

Smagorinsky, J., 1963: "General Circulation Experiments with the Primitive Equations. I: The Basic Experiment," *Mon. Wea. Rev.* 91(3): 99-164.

Thompson, L. M., 1968: "Impact of World Food Needs on American Agriculture," *Jour. Soil and Water Res.* 23.

Thompson, L. M., 1969: "Weather and Technology in the Production of Wheat in the United States," *Jour. Soil and Water Res.* 24: 219-224.

Thompson, L. M., 1970: "Weather and Technology in the Production of Soybeans in the Central United States," *Agr. Journ.* 62: 232-236.

Thompson, L. M., 1966: "Weather Variability and the Need for a Food Reserve," Iowa State University, Center for Agriculture and Economic Development, Report #26, 101 pp.

Webb, Thompson, III and R. A. Bryson, 1972: "Late- and Post-Glacial Climatic Change in the Northern Midwest, U.S.A.: Quantitative Estimates Derived from Fossil Pollen Spectra by Multivariate Statistical Analysis," *Quaternary Res.* 2(1): 70-115.

Wick, Gerald, 1973: "Where Poseidon Courts Aeolus," *New Scientist,* January 18, p. 123.

Wigham, D., and H. Leith, 1971: "Eastern Deciduous Forest Biome Memo Report," #71-9.

Winstanley, Derek, 1973: "Recent Rainfall Trends in

Africa, the Middle East, and India," *Nature* 243: 464-465.

Yamamoto, G., and M. Tanaka, 1972: "Increase of Global Albedo Due to Air Pollution," *J. Atmos. Sci.* 29(8): 1405-1412.

Yin, M. T., 1949: "A Synoptic-Aerologic Study of the Onset of the Summer Monsoon over India and Burma," *Jour. Meteor.* 6: 393-400.

Appendix II

CIA Report: "Potential Implications
of Trends in World Population,
Food Production, and Climate"

CENTRAL INTELLIGENCE AGENCY
Directorate of Intelligence
Office of Political Research

OPR-401
August 1974

Contents

Key Judgments

Trying to provide adequate world food supplies will become a problem of over-riding priority in the years and decades immediately ahead—and a key role in any successful effort must fall to the US. Even in the most favorable circumstances predictable, with increased devotion of scarce resources and technical expertise, the outcome will be doubtful; in the event of adverse changes in climate, the outcome can only be grave.

The momentum of world population growth, especially in the less developed countries (LDCs), is such that even strong measures taken now to reduce fertility would not stop rapid growth for decades. Thus, most LDCs must cope with the needs of much larger populations or face the political and other consequences of rising death rates.

Demand for food rises inexorably with the growth of population and of affluence. Increases in supply are less certain. Man's age-old concerns about the adequacy of food supplies have resumed with particular urgency since the crop-failures of 1972.

The rich countries need have no fear of hunger, though the relative price of food will probably rise at times.

The poor, food-deficit LDCs must produce most of the additional food they will need to support their growing populations. They cannot afford to import it, nor is it likely they can count on getting enough aid from the food-exporting countries. They face

NOTE: *This study was prepared by the Office of Political Research of the Central Intelligence Agency. It does not, however, represent an official CIA position. The views presented represent the best judgments of the issuing office which is aware that the complex issues discussed lend themselves to other interpretations.*

serious political, economic, and cultural obstacles to raising output, however, and are in for considerable strain at the least, and probably for periods of famine.

The US now provides nearly three-fourths of the world's net grain exports and its role is almost certain to grow over the next several decades. The world's increasing dependence on American surpluses portends an increase in US power and influence, especially vis-a-vis the food-deficit poor countries. Indeed, in times of shortage, the US will face difficult choices about how to allocate its surplus between affluent purchasers and the hungry world.

The implications for the world food situation and for US interests would be considerably greater *if* climatologists who believe a cooling trend is underway prove to be right.

If the trend continues for several decades there would almost certainly be an absolute shortage of food. The high-latitude areas, including the USSR and north China, would experience shorter growing seasons and a drop in output. The monsoon-fed lands in Asia and Africa would also be adversely affected.

US production would probably not be hurt much. As custodian of the bulk of the world's exportable grain, the US might regain the primacy in world affairs it held in the immediate post-World War II era.

In the worst case, *if* climate change caused grave shortages of food despite US exports, the potential risks to the US would also rise. There would be increasingly desperate attempts on the part of powerful but hungry nations to get grain any way they could. Massive migrations, sometimes backed by force, would become a live issue and political and economic instability would be widespread.

In the poor and powerless areas, population would have to drop to levels that could be supported. The population "problem" would have solved itself in the most unpleasant fashion.

The Discussion

I. INTRODUCTION

The widespread crop shortfalls in 1972 and the energy and fertilizer crunches in '73 and '74 have raised anew the basic question of whether the production of food can keep pace with demand over the next few decades. Concern about the capability of many of the poorer countries to provide for their growing population is widespread and rising. Major international conferences planned for the second half of this year—*i.e.*, the World Population Conference in August and the World Food Conference in November—will focus on various aspects of this question.

There is, moreover, growing consensus among leading climatologists that the world is undergoing a cooling trend. If it continues, as feared, it could restrict production in both the USSR and China among other states, and could have an enormous impact, not only on the food-population balance, but also on the world balance of power.

This paper briefly reviews present trends and projections for world population and food production under assumptions of "normal" weather, and then essays a necessarily tentative exploration of the ramifications of a cooling climate. A final section addresses the political and other implications for the US of its potential role as the main food exporter in an increasingly hungry world.

II. PEOPLE

World population is growing at ever faster rates. The annual increase in 1930 was about 1.1%; by 1960 it had risen to about 1.7%. It is now around 2.0% and may still be rising. In numbers, these translate to global totals of about two billion in 1930; three billion in 1960, and four billion by early 1975. At current fertility rates, population would total some 7.8 billion by the turn of the century. The UN medium forecast is for about 6.4 billion by 2000 AD; this assumes substantial declines in fertility between now and then. (See tables 1 and 2 in Annex I.)

Population is growing not because birth rates are rising—indeed they are steady or falling in a number of countries—but because death rates, especially infant mortality, have fallen so sharply.*

The increase in population is very unevenly distributed. In most of the developed countries, growth rates are low—the US is currently under 1% a year. But in most of the less developed countries (LDCs) growth rates are well over 2.5%, and in some of the LDCs they approach 3.5% a year. This difference in rates is such that while the peoples of the LDCs (and China) account for about 70% of the world total, by the turn of the century they will represent nearly 80% of the projected total, as illustrated by Figure 1.**

Moreover, the momentum of population growth is so strong in the LDCs, largely because of the size of the group entering or approaching child-bearing age, that even if the developing countries were to adopt strong measures to lower fertility, it is almost certain that the natural rates of population growth would remain high over the subsequent one or two

---

* Birth and death *rates* refer to numbers per hundred total population, and are thus expressed in percentages. *Fertility*, or fertility rates refer to the numbers of children per mother. The *replacement* rate is an average of 2.1 to 2.5 children per family, depending on mortality rates. This would lead to zero population growth only in the longer run. All such terms refer to natural growth and omit the effects of migration.

** China is not otherwise included in the group of LDCs in this review of population and food trends because its population growth is not so rapid; nor does it rely heavily on imported food.

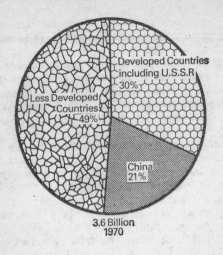

3.6 Billion
1970

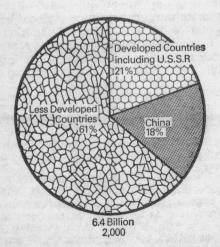

6.4 Billion
2,000

Figure 1. Change in population distribution, 1970 and 2000. (Based on UN "Medium" population projection.)

decades. And, even if fertility fell dramatically and quickly, the number of young couples is so large that population would not stop growing for at least five decades—barring, of course, a major rise in death rates. Without such measures, growth rates would remain high much longer.*

Given the growth rate and age structure of their current populations, it is thus certain that most LDCs will have to provide food for much larger populations before the end of the century or face the political and social consequences of rising death rates. Some will probably have to do both.

All population projections currently in use rest on the assumptions that infant mortality will continue to decline in the LCDs and that life spans will gradually increase, *i.e.,* until they are near those in the developed countries.** Such assumptions in turn depend on some fairly optimistic expectations about improvements in public health and the general nutritional level of the more miserable segments of LDC populations. The projections also imply that wars, famine and plague (while not ruled out) will not be of sufficient magnitude to affect death rates materially on a global basis. For the purposes of this discussion, consideration of major or catastrophic wars is ruled out, but not the other great scourges—famine and plague. These may occur on a scale large enough to raise global death rates if

* If, Mexico, for example, with a current population of about 58 million, reached a replacement rate of fertility by 1980-85, its population would level off at about 110 million by the middle of the next century. If Mexican fertility did not drop to replacement levels until the year 2000, its population would level off at about 170 million in the last half of the next century.

** Under the current medium projections, life expectancy for the world as a whole rises from 59 (about today's level) to 66 by 2000 A.D.: for the LDC's as a group, it is projected to grow from about 54 to 64 over the same period. In large part, closing this gap will come from reductions in infant mortality rates which now range from 1 to 4 per 100 live births in the developed countries to 20 or more in the poor countries.

food shortages become more severe and chronic than they have been in recent years.*

III. FOOD

If population grows as projected, the question of food is two-fold: will there be enough additional output to support the coming billions; and will the distribution systems—physical, administrative, and economic—be adequate to supply food where needed?

Need and Demand

A recent UN Food and Agriculture Organization (FAO) study estimates that 20–25% of the peoples of most countries in Asia and Africa now suffer from serious under-nutrition. Implicit in almost all projections of the coming growth in demand for food are the assumptions that most LDCs will see no real improvement in the average diet over the next decade or two, and that there will be no widespread programs to alleviate malnutrition among the most ill-fed groups. Obviously, the *need* for food is and will continue greater than the *demand* for food, which implies ability to purchase.

Directly consumed cereals provide about half the world's food energy; consumed indirectly in the form of livestock products, they account for much of the rest. And this is where the difference in demand for food by rich and poor countries emerges most forcefully. It requires several pounds of grain to produce a pound of meat. In effect meat eaters consume food

* Famine and plague are closely related in that during periods of famine or poor nutrition, many succumb to diseases they could otherwise survive. Famine often causes the hungry to gather at food distribution points which facilitates the rapid spread of disease. Moreover, if social order including public health services falters, as frequently happens in such conditions, then prevention and treatment of such scourges as cholera and typhoid become less and less possible. For these and other reasons then, the avoidance of epidemics depends heavily on adequate food supplies.

energy twice—first in the form of feed-grain to grow the animal, and then the animal itself.* As income rises, there is an almost universal tendency to eat more protein, especially in the form of animal-products. Annual beef consumption in the US, for example, grew from 55 pounds a person in 1940 to 117 pounds in 1972. In the other industrial countries— Western Europe, Japan, and the USSR—dietary habits resemble those of the US in the 1940s. Since pastures and other natural sources of forage are no longer adequate to feed the animals currently raised for food in most places, increased demand for animal protein requires more grain and other food-stuffs like soybeans.

Forecasts of demand for food over the coming decades rest heavily then on two main factors: the growth of population, especially in the LDCs where demand is most affected by population changes; and income levels in the richer countries where demand is fueled by higher income as much as by population growth.**

Both the US Department of Agriculture (USDA) and the Food and Agriculture Organization (FAO) have recently projected demand for food over the next decade. For the world as a whole they forecast an annual growth of 2.3 to 2.5%. The lower estimate by the USDA is based on the assumption that there will be practically no increase in per capita consump-

---

* In South Asia, for example, most people live on about 400 pounds of grain a year. North Americans, at the other extreme, consume nearly a ton a year, but less than a fifth of it is eaten directly in the form of cereal, bread, etc. The rest is used indirectly as livestock-feed.

** While it is clear that projected demand for food includes protein (in the form of fish, soybeans, pulses, peanuts, etc.) the role of grain as food and feed is so great that this discussion will hereafter focus on grain alone. Protein supplies themselves will be heavily dependent on grain availability since the fish catch has not been growing much in the past 5-6 years; soybeans are not very responsive to fertilizer and hence increased output depends largely on planting more acres (on land that is also suitable for grain). In short, the bulk of protein increases will depend on feed-grains.

tion in the LDCs, and some growth in income and grain imports by the rich countries. The higher FAO forecast is based on more optimistic assessments of economic development in the LDCs where food demand is expected to grow 3.7% a year, which implies some increase in per capita consumption.

Neither forecast can be certain of demand in the developed countries because so much will depend on their policies. This applies with particular force to estimates of the amount of grain that will move in trade between richer countries. For example, will the governments of the USSR and East Europe decide to provide more meat per capita, even if they have to import extra grain, or will they tighten their belts in poor harvest years as they did in the 1960s? Such decisions will have considerable impact on the global demand for grain over the next decades. In any event, it seems plausible to assume that world demand will grow at least 2.3% a year on average and possibly faster.

Supply: The Record

Food production has increased at a fairly steady pace over the past several decades—about equally as fast in the developed and less developed countries. In aggregate terms, the LDCs' performance was impressive. Total food output rose nearly 66% between 1954 and 1973, or more than 2.5% a year. They did even better in grain production: the area planted to grain rose by about 35%, and yield rose nearly as much.

Since population grew vigorously in the LDCs, however, per capita production changed very little: more recently, per capita consumption has probably declined in many LDCs. A number of countries experienced periods of severe malnutrition and growing dependence on food imported from the developed countries.*

---

* USDA estimates indicate a decline in grain consumption per capita in Mexico and Central America, most of the rest of South America except Argentina, Central Africa, and South

The world pattern of grain trade has changed over the same period. Most regions have become net importers, and the volume of such imports has risen, as indicated by the following table.

Net Exports (+) & Imports (−) in million metric tons[1]

	1948-52	1960	1966	72/73[2]	73/74[2]
North America	+23	+39	+59	+89	+92
Latin America	+ 1	0	+ 5	− 3	− 2
Western Europe	−22	−25	−27	−18	−20
East Europe and USSR	0	0	− 4	−26	−12
Africa	0	− 2	− 7	− 1	− 5
Asia	− 6	−17	−34	−38	−49
Australia and New Zealand	+ 3	+ 6	+ 8	+ 7	+ 9

[1]Totals will not balance because of stock changes and rounding.
[2]Different series, but indicative of trend. 73/74 are preliminary.

Twenty years ago, North America exported mainly to Western Europe; most other regions were basically self-sufficient. Now, the whole world has become dependent on North America for grain—feed grains mainly to Europe and Japan, food grains elsewhere. (See Figure 2.) The US now supplies nearly three-fourths of the net global exports, and Canada between 15 and 20%.

In 1972, India's monsoon season was poor, China had drought in the north and floods in the south, the USSR experienced both drought and a short growing season, and drought was particularly oppressive in

East Asia. For LDCs as a group, the USDA estimates annual consumption in kilograms per capita as follows: 1964-66—166, 1969-71—173, 1971-72—168, 1972-73—161, 1973-74—164.

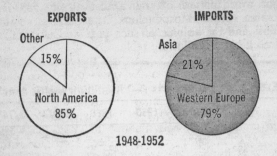

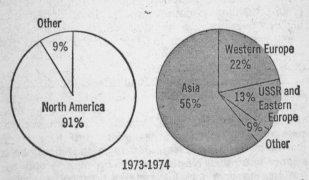

Figure 2. Direction of net grain trade 1948-52 and 1973-74 (prelim.)

parts of Central America and Africa. The results were starvation for some, hunger for many, a rapid rise in food prices everywhere—and a drastic drawdown of existing world stocks of grain.

Supply: Current Prospects

Until very recently, there were two major factors available to cushion the effects of poor harvests: the huge grain stocks in the US (and to a lesser extent in other exporting countries) and the acreage held out of production in the US land bank. Now, stocks are so low they cannot make up for a crop failure in any major area this year. And almost all the US land reserve is back in production. Thus, unless there is exceptionally good weather this year and next, stocks cannot be rebuilt quickly.

Harvests this year and next are therefore of critical importance, especially to the poor food-deficit countries.*For these countries a poor or mediocre harvest could be devastating. Unless, for example, the Indian monsoon improves soon, India will have to import far more grain than it had planned to. Yet, the worldwide inflation of energy, fertilizer, food and other prices has imposed severe strains on its foreign exchange holdings. Mediocre harvests in other major grain-producing countries would drive prices even higher.** In such a situation, there might not be enough surplus to keep India from famine, even if it could afford the cost of imports—or if it somehow got grain on concessional terms.

The current shortage of food affects the developed countries quite differently. For the rich, there are the annoyance and domestic repercussions of high food prices; and for the main importers, an adverse impact on the balance of payments. But there is no fear of real hunger (except perhaps among the poorest segments of their population). The major food exporters—the US and Canada—can expect a sizable boost in earnings from agricultural exports. But if global harvests are poor, these exporting countries will face the difficult choice of whether to sell food to the richer importers or to give it to the poor ones when there is not enough to cover both needs. In short, famine relief on any major scale would have to come from reduced consumption by the well-fed—which would be a difficult and divisive process, even with the best will in the world.

* Not all LDCs fall into this category. Many, especially the oil producers, will be able to afford food imports.
** As indicated by Tables 3-4 in Annex I, less than 10% of the world's harvest constitutes net exports, but relatively small tonnages make a critical difference for food-deficit countries and have a large impact on price.

The Longer Range Outlook

Assuming that the world squeaks by the next few years with good harvests, what of the longer run? As the recent FAO study on the future demand for and supply of food stresses, it is far more difficult to forecast the growth of food production than the rise in demand for it. Here, there is a wide range of opinion among the experts which may be put into two, albeit oversimplified, schools of thought—the "optimistic" and the "pessimistic." They differ far more on their assessments of prospects over the longer run than for the next few years.

The *optimists* stress the theoretical capacity of world agriculture to increase output, by a variety of technical measures, 2–3% a year until at least sometime in the next century.

—On the basis of past performance and improvement in technology, rising demand can be met, provided (1) normal weather prevails (*i.e.*, average conditions similar to the past few decades which cancel out both unusually good and unusually bad years); (2) adequate inputs of fertilizer, pesticides, etc., are available at reasonable prices and (3) prices paid to farmers provide adequate incentives to raise output.

—According to US Department of Agriculture estimates, the US is capable of a 50% increase in feed-grain production by 1985, and thus of providing for almost any foreseeable increase in world import demand for coarse grains (mainly feed grain). Moreover, USDA considers that wheat production could increase by at least a third in the same period. Such gains would come almost entirely from higher yields.

—Under its conservative projections of demand, the USDA foresees cereal production capacity growing faster than consumption: hence the feasibility of rebuilding world reserve stocks, and the possibility of lower prices.

According to the optimists, net grain imports by LDCs would rise from 15.5 million metric tons (1969-71 average) to 40–45 million metric tons in 1985. This assumes that production in these countries could grow slightly faster than 2.6% a year while demand would grow about 3%.

In terms of per capita consumption in the LDCs, these projections imply no appreciable change over this period—only from 181 kilograms a person to 186 by 1985. In short, even the optimistic school of thought projects no real improvement in nutrition for the very poor food-deficit LDCs who make up the bulk of total LDC population. Moreover, many of the poor countries will be unable to pay market prices for the projected level of imports especially if the cost of other essential imports like oil remains high. Thus, unless even the optimistic projections about production in the LDCs are too low, many of the food-deficit LDCs are likely to be in for serious trouble within the next 5–10 years.*

The *pessimists* are dubious about the ability of the world, especially the LDCs, to increase food production at the rates discussed above. They cite a number of very important constraints on ever-rising output.

—Additional arable land is practically unavailable in much of the world, including China, India, Japan, and other parts of Asia. Much of the arable areas of China, for example, are already double and triple cropped. Most additions will require costly capital improvements such as swamp drain-

* FAO projections of food demand and output assume a slightly greater increase in LDC consumption, hence a larger volume of imports. They, too, are quite pessimistic about LDC ability to pay for such imports and hope LDCs can rely on aid. In the 1950's and 60's food aid (much of it US PL 480) amounted to about a third of the total food imports of the LDCs. Such shipments have declined sharply in the past couple of years, however, as the relative cost of food has risen and surplus stocks have been used up.

age, river diversion, complexes of dams and irrigation canals, or construction of desalting plants to make ocean water available. Moreover, the world as a whole is losing several million acres of arable land each year to erosion, salinization, and the spread of cities, industry and roads.*

—Thus, nearly all the increased demand for food will have to be met by higher yields. There are some indications that the rate of growth in LDC output has slowed in recent years: the cheaper and more promising projects have been completed; the most receptive and dynamic farmers have, in many cases, already adopted the new varieties and more modern methods.

—In the developed countries, yield gains may also be slowing. Costs are rising rapidly and it requires ever greater input to achieve an additional unit of output.

—Clearly, the greatest potential for increased food production over the longer run lies in the LDCs, where yields are far below those of the developed countries.** But the social, political and economic obstacles to such development are, in the opinion of the pessimists, formidable.

—Adequate incentives and inputs for farmers imply a major shift in the rural-urban terms of trade in most LDCs. If food prices paid farmers go up, the urban poor cannot afford the increase. Either they get subsidized food, or starve. Few LDCs have been willing to pay adequate subsidies yet, nor is it clear where they could get the necessary funds to do so.

—Moreover, the political commitment to agricul-

---

* Until about 1930, most of the increase in food production came through putting additional land under cultivation and/or having more people at work on the land. Since then, technological improvements—new strains of seed, fertilizer, irrigation, pesticides and mechanization—have played an increasingly important role in raising output.

** Rice yields in India average about ⅓ those of Japan; corn in Thailand and Brazil less than a third of the US.

ture has thus far been lacking. In most LDCs, the governing policy has been either to ignore or to soak the peasants in order to promote industry and keep the city-dweller reasonably content. Reversal of this policy would require enormous inputs of capital and skilled personnel, both in notoriously short supply in most LDCs.

The optimists would argue that as relative food prices rise, as rise they must in the food-deficit LDCs, market forces will call forth increased production. Hopefully they are right on this, but the social and political turmoil implicit in such a price rise is considerable. And large numbers of the very poor would be likely to succumb to famine or hunger-induced diseases long before a new balance were achieved.

In short, whichever school of thought proves out, it seems clear that the world of the poor, at least, will experience continued food shortages and occasional famine over the coming decades. Under either assumption, the developed countries can expect to remain well fed, though perhaps not able to raise their consumption of grain-fed animal products as fast as they might want to. The disparity between the rich and poor is thus likely to get even wider. And the world's dependence on North American agriculture will continue to increase.

IV. CLIMATE *

The precarious outlook for the poor and food-deficit countries, and the enhanced role of North American agriculture in world food trade outlined above were

* Discussion of the nature and impact of possible climate change is, of necessity, highly speculative and therefore controversial. Various experts will disagree with some or many of the implicit assumptions. For example, the Office of Economic Research thinks that too little is known about possible climate change and its potential impact on food production to warrant a discussion of possible adverse implications for food supply.

predicated on the assumption that normal weather will prevail over the next few decades.* But many climatologists warn that this assumption is questionable; some would say that it is almost certainly wrong.

Perhaps the simplest worry is marked variation within the prevailing weather patterns. The US middlewest has had moderate to severe droughts every 20 to 25 years—*e.g.*, 1930s, mid-1950s—as far back as the weather records go. If this pattern holds, the main US granary (now also the mainstay of world grain trade) could expect drought and consequent crop shortfalls within the next several years. If world grain stocks were near today's low levels when this occurred, there would be a severe pinch on world food supplies even if all the other main producing areas had average to good weather.

The extent of this shortage would of course depend on the degree of US drought and its duration. But almost any drop in US output during the next several years would have considerable impact on the price and on the availability of food for the poor foodimporting nations.

Such a cyclical phenomenon, however, would probably last less than a decade. While its impact could be severe, there would be reasonable hope of improvement within a short time.

Far more disturbing is the thesis that the weather we call normal is, in fact, highly abnormal and unusually felicitous in terms of supporting agricultural output. While still unable to explain how or why climate changes, or to predict the extent and duration of change, a number of climatologists are in agreement that the northern hemisphere, at least, is growing cooler.**

---

* Normal weather is defined by climatologists as that which has been experienced in the three previous decades.

** According to Dr. Hubert Lamb—an outstanding British climatologist—22 out of 27 forecasting methods he examined predicted a cooling trend through the remainder of this century. A change of 2°-3°F. in average temperature would have an enormous impact.

—Iceland, because of its location, is a good indicator of changes in the whole Northern Hemisphere. The weather records and evidence for Iceland indicate that the past 4 decades were the most abnormal period in the last 1000 years—much much warmer. (See Figure 3.)

—The arctic ice area has perceptibly increased in the past few years.

—The English growing season has been shortened by a week or more since the 1940's.

The best estimates of climate over the past 1600 years indicate that major shifts have taken place more than a dozen times. The maximum temperature drop usually occurred within 40 years of inception of a cooling trend; and the earliest return to "normal" required 70 years.

A number of meteorological experts are thinking in terms of a return to a climate like that of the 19th century. This would mean that within a relatively few years (probably less than two decades, assuming the cooling trend began in the 1960s) there would be broad belts of excess and deficit rainfall in the middle-latitudes; more frequent failure of the monsoons that dominate the Indian sub-continent, south China and western Africa; shorter growing seasons for Canada, northern Russia and north China. Europe could expect to be cooler and wetter.*

Of the main grain-growing regions, only the US and Argentina would escape adverse effects. In both, the cooler climate at the higher latitudes could be offset by shifting crop-belts equator-ward. In the US, during the 1800's, the mid-west grain areas were cooler and wetter; the southwest was hotter and drier, and the north-east slightly cooler.

Too little is known about the effect of such climate changes on yield to predict the quantitative impact on

---

* For a layman's explanation of one of the more plausible climate theories accounting for such changes, that of Dr. Reid Bryson, see Annex II.

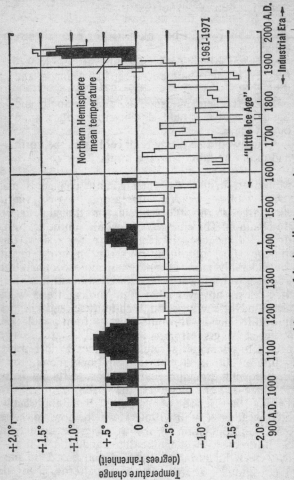

Figure 3. A thousand-year history of Iceland's temperature.

production in the US or in other areas; but some general effects of a major climate change in terms of the global grain output are suggested.

—US output might be unaffected or even slightly enhanced,
—a shorter growing season would restrict production in the high latitude areas, like Canada and the USSR.
—More frequent monsoon failures in South and South-east Asia would significantly reduce grain output there.
—China would be hit by both cooling in the north and monsoon failures in the south.

Moreover, in periods when climate change is underway, violent weather—unseasonal frosts, warm spells, large storms, floods, etc.—is thought to be more common. The change itself would not be smooth, and even if the drop in temperature were slow, the disruptive effect of violent weather on crops might be considerably more adverse than mere cooling. But too little is yet known to be definite about this.

It is clear, however, that *if* a cooling trend were to have adverse effects on high-latitude and on the monsoon-fed lands, it would pose a food-population problem of the gravest nature. Many LDCs are already expected to encounter serious difficulties in increasing agricultural output as fast as their populations grow; more frequent droughts would almost certainly frustrate whatever hope of success they had.

During the period of "normal" (or abnormally good) weather (1930s–1960s) which now may be ending, the population of the world grew more than 50%. Moreover, most major dams and irrigation systems were built during this period and based on prevailing rainfall patterns. If these patterns changed, such systems would be less useful. Most of the hybrids and all of the "green-revolution" strains were developed to use the warmth and moisture prevailing in this period. Significant change in temperature or rainfall pattern could negate most of these advances in yield. Experts are too uncertain about the possible magnitude of the cooling trend and change in rainfall to be able to chart the effects on irrigation systems and hybrids, but production in many countries almost certainly would be cut.

Clearly, agronomists could develop new strains more suitable to different weather and every effort would be made to counter the adverse effects of a climate change. New methods of manufacturing food-stuffs and stretching what was available would help, *e.g.,* texturized vegetable protein, milk made directly from grass. Unconventional food sources, like the yeasts that can be made to grow on petroleum, would also be tried. But in most cases the cost of such supplemental foods is still greater than natural foods and would thus be of little help to the poorest and neediest groups for a decade or more.

The USSR, China, and South Asia would probably need large imports. How much—the most critical question—would depend heavily on how far and how fast the climate patterns changed. If the cooling trend were marked and persistent, then a physical shortage of food would seem inevitable. That is, no matter what the price or the distribution arrangements, there would not be enough produced to feed the world's population—unless the affluent nations made a quick and drastic cut in their consumption of grain-fed animals. Even then there might not be enough.

V. POLITICAL AND OTHER IMPLICATIONS

With or without "normal" weather, the US is almost certain to increase its dominance of the world's grain trade over the next couple of decades. This enhanced role as supplier of food will provide additional levers of influence, but at the same time will pose difficult choices and possibly new problems for the US. The magnitude and range of implications differ radically depending on whether "normal" or much cooler weather is postulated.

Assuming Normal Weather

The growing dependence of poor food-deficit LDCs on imported grain and the continued desire of affluent peoples to increase their consumption of animal products promise generally strong markets for US grain exports and considerable benefits to the US balance of payments. Moreover, ability to provide relief food in periods of shortage or famine will enhance US influence in the recipient countries, at least for a time.

This dependence is also likely to lead to resentment of the US role on the part of the dependent countries. Nevertheless, many will find it expedient to accommodate US wishes on a variety of issues. Others, perhaps with the backing of the USSR or China, may seek to establish some international controls over the allocation of world food stocks.

The US, for its part, will face the difficult and recurring issue of where its grain should go. In times of shortage when food prices rise, it will be hard to decide how much should be reserved for domestic consumption, how much should be sold at high prices, and how much should be given in aid to the needy. Each decision will have domestic and balance of payments repercussions, and will engage the humanitarian impulses of the country. Moreover, it may be difficult to choose among LDCs as recipients. Whatever the choices, the US will become a whipping boy among those who consider themselves left out or given only short shrift. The few other nations which might have some surplus will be tempted to use it for their own political ends.

There will probably be a number of times when there is not enough surplus: for example, a repeat of the crop shortfalls of 1972 when a number of major agricultural areas simultaneously experienced bad weather. If food stocks were low at the time, there would not be enough to supply the food-deficit LDCs and the feed-grain needs of the affluent countries. The very poor LDCs would almost certainly be unable to pay for the necessary imports and, even with aid, many would face famine.

The elites of many LDCs tend to regard periodic famine as either natural or at least beyond their power to prevent, *e.g.,* Bihar in 1967, Ethiopia and the Sahelian states in 1973. But the rural masses may become less docile in the future and if famine also threatens the cities and reduces the living standards of the middle classes, it could lead to social and political upheavals which cripple governmental authority. The beleaguered governments could become more difficult to deal with on international issues either because of a collapse in ability to meet commitments or through a greatly heightened nationalism and aggressiveness.

Developed countries that import large amounts of grain need not fear hunger or real privation, but will experience additional financial strain in years when crops are mediocre or poor. In such times, a sharp rise in food prices will affect their living standards. It has been decades since any rich country has been short of food and such an unfamiliar situation could generate great social and political stress. The US, as potential supplier, would gain influence; it might also be blamed for part of the shortage.

In sum, if the weather is "normal", it is essentially the poorer LDCs that would become ever more dependent on US food exports. Many will be unable to increase their own exports enough to pay for such imports and will either have to get food on concessional terms or face increasing shortages including a degree of famine which applies a Malthusian check on their population growth. This will pose problems for the US and other rich countries, particularly if the affected LDCs decline into an ungovernable state of confusion.

Assuming the Cooling Trend Continues

A discussion of the implications of a cooler climate within the next several decades must, necessarily, be highly speculative. Yet even tentative assessments of the potential prospects may be useful in view of the evidence of cooling that now exists. One obstacle to more definite consideration is that climatologists are not able to predict how far the cooling trend might go. In the next 5 paragraphs, the discussion is based on the assumption of cooling great enough to cut the production of higher latitude areas (Canada, the USSR, and north China) and increase the frequency of drought in the monsoon-fed countries. It further assumes that US surpluses would still cover most needs, except in bad years. The last two paragraphs assume that the regression in weather reaches the point where even the best efforts of the US would not normally be sufficient to meet the minimum needs of the major food-deficit areas.

In a cooler and therefore hungrier world, the US's near-monopoly position as food exporter would have an enormous, though not easily definable, impact on international relations. It could give the US a measure of power it had never had before—possibly an economic and political dominance greater than that of the immediate post-World War II years.

A substantially cooler climate could add new and powerful countries to the list of major importers and reduce Canada's exportable surplus. Among the most immediate effects would be rapid increases in the price of food in almost all countries, which would create internal dislocations and discontent. The poor, within countries and as national entities, would be hardest hit. What is happening now to the poor in India and in drought-stricken Africa is probably a pale sample of what the food-deficit areas might then experience.

In many LDCs, the death rate from malnutrition and related diseases would rise and population growth slow down or cease. Elsewhere, there might be waves of migration of the hungry towards areas thought to have enough food. The outlook then would be for more political and economic instability in most poor countries as well as for growing lack of confidence in leaders unable to solve so basic a problem as providing food.

For the richer countries, the impact would be mitigated, at least, by their very wealth. While standards of living in countries needing to import large quantities of food would probably decline, there would be little danger of starvation. Nevertheless, there would be varying degrees of economic dislocation and political dissatisfaction whose results would be very difficult to forecast.

In bad years, when the US could not meet the demand for food of most would-be importers, Washington would acquire virtual life and death power over the fate of multitudes of the needy. Without indulging in blackmail in any sense, the US would gain extraordinary political and economic influence. For not only the poor LDCs but also the major powers would be at least partially dependent on food imports from the US.

In the worst case, where climate change caused grave shortages of food despite US exports, the potential risks to the US would rise. There would be increasingly desperate attempts on the part of the militarily powerful, but nonetheless hungry, nations to get more grain any way they could. Massive migration backed by force would become a very live issue. Nuclear blackmail is not inconceivable. More likely, perhaps, would be ill-conceived efforts to undertake drastic cures which might be worse than the disease; *e.g.*, efforts to change the climate by trying to melt the arctic ice-cap.

In the poor and powerless areas, population would have to drop to levels that could be supported. Food subsidies and external aid, however generous the donors might be, would be inadequate. Unless or until the climate improved and agricultural techniques changed sufficiently, population levels now projected for the LDCs could not be reached. The population "problem" would have solved itself in the most unpleasant fashion.

The potential implications of a changed climate for the food-population balance and for the world balance of power thus could be enormous. They would become far clearer and possibly more manageable if the extent of possible cooling were thoroughly investigated and if the potential impact of that cooling were quantified.

Annex I

TABLES ON POPULATION GROWTH AND ON WORLD GRAIN PRODUCTION AND TRADE

Table 1
World Population Projections

	Assuming Constant Fertility		Assuming Declining Fertility*	
	Millions	Growth Rate**	Millions	Growth Rate
1970	3,600	....	3,600	....
1985	5,200	2.4%	4,858	2.0%
2000	7,822	2.8%	6,407	1.9%

*UN "medium" projection.
**Annual average rate since preceding date.

Table 2

Regional Distribution and Growth Rates of Population: 1970–2000

(UN projections, medium variant)

	1970	1985	2000	Annual Avg. Growth Rate 1970-2000
	Millions			Percent
Developed Countries	1,084	1,234	1,368	0.8
Western	736	835	920	0.7
Of which				
US	205	236	264	0.9
Communist	348	399	447	0.8
Of which				
USSR	243	283	321	0.9
Less Developed Countries	2,537	3,624	5,039	2.3
Communist	794	1,007	1,201	1.4
Of which				
China	758	955	1,127	1.3
Other	1,743	2,616	3,838	2.7
Far East	49	66	83	1.8
Other Asia	1,090	1,625	2,341	2.6
Africa	352	536	834	2.9
Latin America	248	384	572	2.8
World Total	3,621	4,858	6,407	1.9

Table 3

Total Grains (wheat, coarse grain, and milled rice): Production, Disappearance, and Net Trade
(Million metric tons)

Region or country	1964/65—1966/67			1969/70—1971/72		
	Prod.	Disap.	Net trade[1]	Prod.	Disap.	Net trade[1]
Developed						
United States	175.1	144.7	43.3	208.7	168.3	39.3
Canada	33.1	18.1	14.3	34.8	22.4	14.7
EC-9	79.6	100.0	—20.0	93.2	110.4	—16.6
Other Western Europe.	22.8	28.2	— 5.8	29.0	33.9	— 5.0
Japan	14.0	23.9	—10.3	12.7	28.2	—14.4
Australia and New Zealand	13.5	5.3	7.0	14.9	6.0	10.8
South Africa	7.9	5.6	0	10.1	6.9	1.2
Total	346.0	325.8	28.5	403.4	376.1	30.0
Communist						
East Europe	64.8	71.7	— 7.0	75.0	82.2	— 7.3
USSR	139.6	139.9	— .2	168.8	164.9	4.0
China	139.6	144.2	— 4.5	159.3	162.4	— 3.1
Total	344.1	355.8	—11.7	403.1	409.5	— 6.4
Less developed						
Mexico and Central America ...	13.7	14.3	— .5	16.0	18.0	— 1.8
Brazil	17.0	18.1	— 1.9	20.6	21.7	— .7
Argentina	17.7	8.7	10.2	19.3	10.7	8.3
Other South America..	7.3	9.0	— 1.8	7.6	10.8	— 3.1
North Africa	11.6	14.8	— 3.2	15.0	18.3	— 3.3
Central Africa	27.7	28.3	— .7	30.7	31.8	— 1.0
West Asia	25.9	28.4	— 2.5	29.4	34.2	— 5.1
South Asia	86.0	96.5	—10.1	114.1	118.0	— 4.7
Southeast Asia	30.6	16.3	3.4	25.6	22.0	3.6
East Asia, Pacific	24.5	27.8	— 3.4	31.2	38.4	— 7.7
Total	262.0	262.2	—10.5	309.5	323.9	—15.5
World total	952.1	943.8	6.3	1,116.0	1,109.5	8.1

[1] Minus indicates net trade imports. Totals may not add due to rounding or stock changes.

[2] Preliminary.

[3] Projected.

SOURCE: USDA, **World Agricultural Situation**, December 1973.

1971/72			1972/73 [2]			1973/74 [3]		
Prod.	Disap.	Net trade [1]	Prod.	Disap.	Net trade [1]	Prod.	Disap.	Net trade [1]
236.5	174.2	40.0	226.8	179.5	70.1	239.2	177.7	74.4
38.8	23.5	18.4	35.6	23.6	18.7	37.6	22.9	17.1
100.4	112.0	−14.3	103.1	117.4	−13.1	105.4	118.4	−12.5
32.6	35.7	− 4.6	30.2	36.5	− 5.2	28.8	36.5	− 7.4
10.9	28.8	−15.0	11.5	29.4	−16.8	11.6	30.2	−18.8
15.4	6.8	12.1	11.0	6.7	7.3	17.3	6.7	8.8
11.7	7.3	2.9	6.3	7.6	3.4	10.2	7.8	.3
446.3	388.3	39.5	424.5	400.7	64.4	450.1	400.2	61.9
82.1	90.1	− 8.6	87.4	94.4	− 6.6	89.2	95.4	− 6.3
173.7	175.1	− 5.8	160.1	179.1	−19.0	204.2	209.4	− 5.2
164.0	166.3	− 2.3	160.6	165.9	− 5.3	168.0	176.3	− 8.1
419.8	431.5	−16.7	408.1	439.4	−30.9	461.4	481.1	−19.6
16.8	19.4	− 2.1	14.8	19.2	− 4.0	16.3	20.4	− 4.0
20.0	23.3	− .8	19.4	22.4	− 2.7	21.9	24.0	− 2.4
15.5	10.3	7.6	22.2	11.3	7.6	20.8	12.0	9.2
7.4	11.4	− 3.8	7.2	11.5	− 4.3	7.2	12.4	− 5.1
15.6	18.8	− 3.3	16.4	19.5	− 3.2	15.0	19.1	− 4.3
30.9	31.9	− 1.2	29.2	30.2	− 1.0	27.0	28.3	− 1.3
30.5	33.8	− 6.6	32.1	35.8	− 3.4	28.1	35.4	− 6.1
114.4	115.5	− 3.5	109.9	121.0	− 4.4	122.0	131.2	− 8.5
25.0	21.1	4.2	21.3	20.6	1.6	25.7	22.5	3.1
32.4	40.1	− 8.9	30.9	42.5	− 9.9	34.0	45.5	−10.9
308.5	325.6	−18.4	303.4	334.0	−23.7	318.0	350.8	−30.3
1,174.6	1,145.4	4.4	1,136.0	1,174.1	9.8	1,229.5	1,232.1	12.0

Table 4

Production, Disappearance, and Net Trade in Grain
(Wheat, coarse grain, and milled rice)

Country and region	1969—71 Prod.	Disap.	Net trade*	1985 Projection Prod.	Disap.	Net trade*
			Million metric tons			
Developed						
United States	208.7	168.3	39.3	286.0	232.4	53.7
Canada	34.8	22.4	14.7	46.0	26.1	19.9
EC-9	93.2	110.4	—16.6	133.5	135.2	—1.7
Other Western Europe	29.0	3.9	—5.0	37.8	44.0	—6.2
Japan	12.7	28.2	—14.4	11.5	46.6	—35.1
Australia and New Zealand	14.9	6.0	10.8	22.7	8.5	14.1
South Africa	10.1	6.9	1.2	14.9	10.8	4.1
Total	403.4	376.1	30.0	552.4	503.6	48.8
Communist						
East Europe	75.0	82.2	—7.3	102.9	104.0	—1.2
USSR	168.8	164.9	4.0	227.3	227.6	—0.3
China	159.3	162.4	—3.1	209.7	214.0	—4.2
Total	403.1	409.5	—6.4	539.9	545.6	—5.7
Less developed						
Mexico and Central America ..	16.0	18.0	—1.8	25.3	30.2	—4.9
Brazil	20.6	21.7	—0.7	31.3	33.3	—2.0
Argentina	19.3	10.7	8.3	25.6	12.9	12.6
Other South America.	7.6	10.8	—3.1	9.6	17.0	—6.9
North Africa	15.0	18.3	—3.3	23.0	33.3	—10.3
Central Africa	30.7	31.8	—1.0	40.2	44.6	—4.4
West Asia	29.4	34.2	—5.1	36.1	47.3	—11.2
South Asia	114.1	118.0	—4.7	180.0	188.8	—8.7
Southeast Asia	25.6	22.0	3.6	40.8	34.6	6.1
East Asia, Pacific ..	31.2	38.4	—7.7	47.6	60.9	—13.2
Total	309.5	323.9	—15.5	459.5	502.9	—42.9
World total	1,116.0	1,109.5	8.1	1,551.8	1,552.1	0.2

*Net trade may not equal the difference between production and disappearance because of stock changes. Minus indicates net imports.

SOURCE: USDA, **World Agricultural Situation**, December 1973.

Annex II

CLIMATE THEORY *

Professor Reid A. Bryson of the University of Wisconsin at Madison has developed a theory and begun to develop a weather forecasting methodology based upon it. He contends that the world is at the end of a golden era: that of benign climate and food surpluses. Moreover, climate change has set in, and it will be 40 to 60 years from now at a minimum (possibly centuries) before we can hope for equally benign weather. He reasons along the following lines.

The earth's atmosphere, hence its weather, is driven by the heat of the sun. Temperature differences—between pole and equator and between surface and upper air—constitute the main working parts of this heat engine and are responsible for pressure differences and the consequent flow of air masses.

At the beginning of this century, temperatures were rising and the mound of cold air that covers the pole was contracting.** By about 1940 it had reached a relatively small size and more of the earth was dominated by warm air from tropical regions. But since

* This is a layman's review for laymen of one of the many theoretical descriptions of climate change that observers agree is occurring. For a fuller explanation of this and other theses, see the forthcoming paper from ORD, "A Study of Climatological Research."

** These phenomena are of greater impact in the northern hemisphere for 2 reasons: there are many more people and much more land mass in that hemisphere.

231

about 1940, the earth has in fact cooled and the polar air mass has expanded.

Why has the earth cooled? There are three main factors involved affecting how much sunlight reaches the earth and how much is re-radiated into space: volcanic dust, man-made dust, and carbon dioxide. The transparency of the atmosphere to incoming sunlight and heat is affected by dust. The main variable sources of dust are volcanic activity and man-made pollution. In the early part of this century, volcanic activity decreased markedly and this increased transparency; temperatures rose.

Bryson estimates that transparency was little affected by man's activities until about 1930 but that since then, man-caused dust has increased rapidly.* And, since the mid-1950's volcanic activity has again become important.

According to his theory, the earth would have cooled due to this dust even more than it has if it had not been for measurable and increasing amounts of carbon dioxide which man has put into the atmosphere by burning fuel (the greenhouse effect).

Temperature changes, though important, are less so than the circulation pattern they help engineer. And here, dust cools the polar regions proportionally more than the tropics. This increases the temperature difference between the two (the temperature gradient). Increased carbon dioxide warms the surface but not the upper atmosphere; this increases the vertical temperature gradient. These two gradients are the main factors that control atmospheric circulation and rainfall.

The net effect is cooler weather in the high latitudes (near the arctic air masses) and a shift southward of the subtropical high pressure areas. The latter control the northward movement of monsoon rain (in the northern hemisphere). This shift, in Bryson's opin-

---

* Man-made dust includes wind erosion of soils left unprotected as well as the more familiar industrial pollutants.

ion, accounts for the prolonged drought in the African Sahelian region.*

North (or poleward) of this sub-tropical high pressure belt, the general effect of temperature gradient changes seems to be that areas dependent on westerlies for rain will have less strong westerlies and hence less rain. Inland on the Eurasian and other large land masses, north-south swings of the polar front (the edge of the great polar air mass) will tend to dominate the weather picture more than in the recent warm period.

Bryson expects a return to something like the climate of the last century, in which case the following tendencies could be expected:

—more rain in the northern half of the US; drier in Central Gulf Coast, southwest, and the northern Rockies. The winter wheat area and the range lands of the high plains would be much wetter. On balance, these changes would probably not affect US food production very much.

—a higher frequency of drought in India, as the northern limits of the monsoon are pushed southward. Perhaps as much as severe drought every three to four years in northern and northwest India.

—persistent drought in Sahelian Africa so long as the sub-tropic high pressure area stays where it is.

—shorter growing seasons in Canada, northern Europe, northern Russia and north China and consequent reduction in grain output.

—more frequent monsoon failures in South East Asia and the Philippines.

* The main temperature change seems to be in the summer, where mean arctic temperatures have dropped 0.5°C. (nearly 1°F.) giving an increased pole to equator gradient of 0.1°C. (almost 0.2°F.) per 1000 kilometers. This would, according to the theory, lower the latitude of the sub-tropic high by more than 30 miles.

Control over volcanic activity is well beyond human capability. Nor does it seem likely that human societies could change their activity patterns so as to reduce the amount of man-made dust which now accounts for perhaps 20–30% of the total. Thus, if the theory is correct, it is reasonable to assume continued cooling and change in weather patterns unless or until volcanic activity slows again. Even then, man-made dust would remain as an important cooling factor.